Python 编程练习与解答

[加拿大] 本·斯蒂芬森(Ben Stephenson)　著
孙鸿飞　史苇杭　译

清华大学出版社

北　京

北京市版权局著作权合同登记号　图字：01-2020-7490

First published in English under the title The Python Workbook: A Brief Introduction With Exercises And Solutions, Second Edition by Ben Stephenson. Copyright © 2019.
This edition has been translated and published under licence from Springer Nature Switzerland AG.
All Rights Reserved.

此版本仅限在中华人民共和国境内(不包括中国香港、澳门特别行政区和台湾地区)销售。未经出版者预先书面许可，不得以任何方式复制或抄袭本书的任何部分。

本书封面贴有清华大学出版社防伪标签，无标签者不得销售。
版权所有，侵权必究。举报：010-62782989，beiqinquan@tup.tsinghua.edu.cn。

图书在版编目(CIP)数据

Python编程练习与解答 /(加)本·斯蒂芬森(Ben Stephenson) 著；孙鸿飞，史苇杭译. —北京：清华大学出版社，2021.1（2024.11重印）
书名原文：The Python Workbook: A Brief Introduction With Exercises And Solutions, Second Edition
ISBN 978-7-302-56955-8

Ⅰ.①P…　Ⅱ.①本…②孙…③史…　Ⅲ.①软件工具－程序设计　Ⅳ.①TP311.561

中国版本图书馆 CIP 数据核字(2020)第 228230 号

责任编辑：王　军　韩宏志
封面设计：孔祥峰
版式设计：思创景点
责任校对：成凤进
责任印制：沈　露

出版发行：清华大学出版社
　　　　　网　　　址：https://www.tup.com.cn, https://www.wqxuetang.com
　　　　　地　　　址：北京清华大学学研大厦 A 座　邮　　编：100084
　　　　　社 总 机：010-83470000　　　　　　　　　邮　　购：010-62786544
　　　　　投稿与读者服务：010-62776969, c-service@tup.tsinghua.edu.cn
　　　　　质 量 反 馈：010-62772015, zhiliang@tup.tsinghua.edu.cn
印 装 者：三河市东方印刷有限公司
经　　销：全国新华书店
开　　本：148mm×210mm　　　印　张：8.125　　　字　数：211千字
版　　次：2021 年 1 月第 1 版　　印　次：2024 年 11月第 6 次印刷
定　　价：49.80元

产品编号：086757-01

译者序

自从 20 世纪 90 年代初 Python 语言诞生至今,已被逐渐广泛应用于系统管理任务的处理和 Web 编程。由于 Python 语言的简洁性、易读性以及可扩展性,在国外用 Python 从事科学计算的研究机构日益增多,一些知名大学已经采用 Python 来讲授程序设计课程。例如卡耐基·梅隆大学的编程基础、麻省理工学院的计算机科学及编程导论就使用 Python 语言讲授。

Python 是一种解释型脚本语言,可应用于以下领域:Web 和 Internet 开发、科学计算和统计、教育、桌面界面开发、软件开发、后端开发等。

学习一种编程语言,首先要找一款合用的集成开发工具,这似乎是自然而然的想法。Python 是一种计算机程序设计语言,是一种面向对象的动态类型语言。IDLE 是 Python 软件包自带的一个集成开发环境,初学者可以利用它方便地创建、运行、测试和调试 Python 程序。建议初学者使用 IDLE 编辑器而非 IDE 的好处在于:

- 专注于 Python 本身,而不是被工具使用问题所困扰。
- 手动运行代码,可以更直观、更深刻地了解脚本的解释执行过程。

- 手动调试代码，有助于代码优化，提高代码把控能力。

解释型语言的优势就是可以写一句执行一句，想到哪儿写到哪儿，不必像编译型语言那样需要将程序全部写完，编译成功后才能运行。使用 Python 的 IDLE，可以验证代码的写法是否正确，查看模块是否安装成功以及确认版本号。IDLE 支持 Tab 键自动补齐，还可供查看某个对象的方法和属性。

很多 Python 初学者热衷于观看各种网络教程、视频教程，甚至还做了很多笔记。经过了长时间的学习之后，发现自己还是无法驾驭 Python。出现这个问题的原因，不是因为他们懂得太少，而是因为知道得太多了——准确地说，是在初级阶段学习了高级阶段的课程。

学习是一个循序渐进的过程，而编程又是一门实践性很强的艺术，因此学习编程需要不断地重复"学习-实践"。真正的实践，就是写一段让自己满意的代码，实现一个独立功能。比如，初学者可以实现从一个文本文件读出内容，做些特别处理，再写入另一个文件，或者写一些算法函数等。

本书的大部分内容是练习题及其解答，只有少数内容用来简要介绍完成它们所需的概念。本书包含 186 个练习，涉及各个学科和日常生活情况。它们可以仅使用大多数介绍性 Python 编程课程中涉及的内容来解决。你完成的每个练习都将增强你对 Python 编程语言的理解，提高你应对后续编程挑战的能力。这些练习与其他学科和日常生活有紧密的联系，希望在读者完成它们时保持对 Python 的兴趣。

本书有多种用途，简明介绍了主要的 Python 编程概念。可将本书作为介绍性编程课程的教科书，也可作为练习题有限的其他教科书的补充。比较自律的人可以只用本书自学 Python 编程。

这里要感谢清华大学出版社的编辑，他们为本书的出版投入了巨大的热情并付出了很多心血。没有他们的帮助和鼓励，本书不可能顺利付梓。

对于这本经典之作，译者本着"诚惶诚恐"的态度，在翻译过程中力求"信、达、雅"，但是鉴于译者水平有限，失误在所难免，如有任何意见和建议，请不吝指正。

译　者

前言

我相信计算机编程是一种最好通过实践来学习的技能。虽然阅读课本上有关程序设计的知识、观看教师在讲堂上创建程序的过程是有价值的，但花时间解决问题，将程序设计的概念付诸实践更为重要。请记住这一点：本书的大部分内容是练习题及其解答，只有少数内容用来简要介绍完成它们所需的知识。

本书包含 186 个练习，涉及各个学科和日常生活情况。它们可以仅使用大多数介绍性 Python 编程课程中涉及的内容来解决。完成的每个练习都将增强你对 Python 编程语言的理解，提高你应对后续编程挑战的能力。这些练习与其他学科和日常生活有紧密的联系，希望在你完成它们时保持对 Python 的兴趣。

本书下半部分提供了大约一半练习的解答。大多数解答都包含简短的注释，解释用于解题的技术或强调 Python 语法的特定知识点。

请读者花时间将自己的解答与本书提供的解答进行比较，即使是在没有遇到任何问题的情况下得出了答案。这种比较可能会暴露出程序中的某个缺陷，或者帮助你熟悉某种技术，若使用更合理的技术，问题可能更容易解决。某些情况下，还可能表明，读者找到了比本书更快或更简单的解题方法。如果在某个练习上卡住了，浏览一下本书的解答也许能帮助解决问题，继续取得进展，而不需要别人的帮助。最后，本书提供的解答演示了良好的编程风格，包括适当的注释、有意义的变量名和尽量少用魔术数字。读者在创建解题程序时，最好使用良好的编程风格，以便计算出正确的结果，使解答清晰、易于理解，并能在将来更新。

包含解答的练习在其名称旁边清楚标记"有解答"。本书的每个练习还标明了示例解答的长度。虽然不应该期望自己的解答长度与示例解答完全匹配，但提供此信息是希望防止读者在寻求帮助之前误入歧途。

本书有多种用途，简明介绍了主要的 Python 编程概念(在本版中是新增的)。可以把本书作为介绍性编程课程的教科书，也可把本书作为练习题有限的其他教科书的补充。比较自律的人可以只用本书自学 Python 编程。本书对每个主题的简明介绍只涵盖了它们最重要的方面，而没有涉及特殊情况或不寻常的情况。无论在本书中使用什么其他资源(如果有)，阅读章节、完成练习并研究所提供的解答都将增强编程能力。

致谢

我要感谢 Tom Jenkyns 博士在本书出版时给出的评论。我根据他提出的有益意见和建议进行了许多改进和更正，提高了本书的质量。

2019 年 3 月，作于加拿大卡尔加里市
Ben Stephenson

目录

第 I 部分 练习

第 1 章 编程概论 ·················· 3
1.1 存储和操作值 ·················· 4
1.2 调用函数 ·················· 5
    1.2.1 阅读输入 ·················· 6
    1.2.2 显示输出 ·················· 7
    1.2.3 导入其他函数 ·················· 8
1.3 注释 ·················· 9
1.4 值的格式化 ·················· 9
1.5 处理字符串 ·················· 12
1.6 练习 ·················· 13

第 2 章 决策 ·················· 27
2.1 if 语句 ·················· 27
2.2 if - else 语句 ·················· 29

- 2.3 if-elif-else 语句 ⋯⋯ 30
- 2.4 if-elif 语句 ⋯⋯ 32
- 2.5 嵌套的 if 语句 ⋯⋯ 32
- 2.6 布尔逻辑 ⋯⋯ 33
- 2.7 练习 ⋯⋯ 35

第 3 章 循环 ⋯⋯ 51
- 3.1 while 循环 ⋯⋯ 51
- 3.2 for 循环 ⋯⋯ 53
- 3.3 嵌套循环 ⋯⋯ 55
- 3.4 练习 ⋯⋯ 56

第 4 章 函数 ⋯⋯ 69
- 4.1 带参数的函数 ⋯⋯ 70
- 4.2 函数中的变量 ⋯⋯ 73
- 4.3 返回值 ⋯⋯ 73
- 4.4 将函数导入其他程序 ⋯⋯ 76
- 4.5 练习 ⋯⋯ 77

第 5 章 列表 ⋯⋯ 87
- 5.1 访问单个元素 ⋯⋯ 88
- 5.2 循环和列表 ⋯⋯ 88
- 5.3 其他列表操作 ⋯⋯ 91
  - 5.3.1 向列表中添加元素 ⋯⋯ 92
  - 5.3.2 从列表中删除元素 ⋯⋯ 92
  - 5.3.3 重新排列列表中的元素 ⋯⋯ 93
  - 5.3.4 搜索列表 ⋯⋯ 94
- 5.4 列表作为返回值和参数 ⋯⋯ 95
- 5.5 练习 ⋯⋯ 96

第 6 章 字典 ·············· 111
- 6.1 访问、修改和添加值 ·············· 112
- 6.2 删除键值对 ·············· 113
- 6.3 其他字典操作 ·············· 114
- 6.4 循环和字典 ·············· 114
- 6.5 字典作为参数和返回值 ·············· 116
- 6.6 练习 ·············· 117

第 7 章 文件和异常 ·············· 125
- 7.1 打开文件 ·············· 126
- 7.2 从文件中读取输入 ·············· 126
- 7.3 行结束符 ·············· 128
- 7.4 将输出写入文件 ·············· 130
- 7.5 命令行参数 ·············· 131
- 7.6 异常 ·············· 133
- 7.7 练习 ·············· 135

第 8 章 递归 ·············· 147
- 8.1 整数求和 ·············· 148
- 8.2 斐波那契数 ·············· 149
- 8.3 计算字符个数 ·············· 152
- 8.4 练习 ·············· 153

第 II 部分 答案

第 9 章 "编程概论"练习答案 ·············· 165

第 10 章 "决策"练习答案 ·············· 175

第 11 章 "循环"练习答案 ·············· 187

第 12 章 "函数"练习答案 ·· 195

第 13 章 "列表"练习答案 ·· 209

第 14 章 "字典"练习答案 ·· 221

第 15 章 "文件和异常"练习答案 ······································ 227

第 16 章 "递归"练习答案 ·· 239

第1部分

练 习

第1章 编程概论

计算机帮助我们完成许多不同的任务。它们也允许我们读新闻、看视频、玩游戏、写书、购买商品和服务、进行复杂的数学分析、与朋友和家人交流等。所有这些任务都需要用户提供输入，比如点击要观看的视频，或者输入应该包含在书中的句子。作为响应，计算机生成输出，如打印一本书、播放声音或在屏幕上显示文本和图像。

考虑一下前一段中的例子。计算机如何知道要请求什么输入？如何知道应该采取什么操作来响应输入？如何知道要生成什么输出，以及应该以什么形式呈现输出？所有这些问题的答案都是"一个人给计算机下达指令，计算机执行指令"。

算法是解决问题的有效且有限的步骤序列。如果一个步骤是明确的、可以执行，它就是有效的。步骤数量必须是有限的(而不是无限的)，这样才能完成所有步骤。菜谱、家具或玩具的装配说明，以及打开密码锁所需的步骤，都是我们在日常生活中遇到的算法例子。

算法的形式十分灵活,可根据算法解决的问题进行调整。单词、数字、线条、箭头、图片和其他符号都可用来表达必须执行的步骤。虽然算法采用的形式各不相同,但所有算法都描述了成功完成任务需要遵循的步骤。

计算机程序是控制计算机行为的指令序列。这些指令告诉计算机何时执行诸如读取输入和显示结果的任务,以及如何转换和操作值以获得期望的结果。在计算机用来解决问题之前,算法必须翻译成计算机程序。翻译过程称为编写程序,执行翻译的人称为程序员。

计算机程序是用计算机编程语言编写的。程序语言有精确的语法规则,必须严格遵守。如果不这样做,将导致计算机报告错误,而不是执行程序员的指令。人们创造了各种不同的语言,每种语言都有其优缺点。目前流行的编程语言包括 Java、C++、JavaScript、PHP、C#和 Python 等。虽然这些语言之间有很大的差异,但它们都允许程序员控制计算机的行为。

本书使用 Python 编程语言,因为它对于新程序员来说易于学习,而且可用来解决各种问题。下面几节将描述从用户读取键盘输入、执行计算和生成文本输出的 Python 语句。后续章节将描述额外的编程语言结构,它们可用来解决更大、更复杂的问题。

1.1 存储和操作值

变量是计算机内存中保存值的指定位置。在 Python 中,变量名必须以字母或下画线开头,然后是字母、下画线和数字的组合[1]。用赋值语句创建变量。要创建的变量名放在赋值操作符(用=表示)的左边,存储在变量中的值放在赋值操作符的右边。例如,下面的语句创建了一个名为 x 的变量,并在其中存储 5。

---

[1] 变量名区分大小写。因此,count、Count 和 COUNT 是不同的变量名,尽管它们很相似。

```
x = 5
```

赋值语句的右侧可以是任意复杂的计算式，包括圆括号、数学运算符、数字和变量等。Python 提供的常见数学运算符包括加法(+)、减法(-)、乘法(*)、除法(/)和求幂(**)。运算符还提供了取整(floor)除法(//)和取模(%)。取整除法运算符计算两数相除所得的商的向下取整值，而取模运算符计算两数相除所得的余数。

下面的赋值语句计算 $1 + x^2$ 的值，并将其存储在一个名为 y 的新变量中。

```
y = 1 + x ** 2
```

Python 遵循数学运算符的通常操作顺序规则。由于 x 是 5(来自前面的赋值语句)，指数运算的优先级比加法更高，赋值运算符右边的表达式的计算结果是 26。然后将这个值存储在 y 中。

相同的变量可以出现在赋值运算符的两边。例如：

```
y = y - 6
```

读者的第一反应可能是：这样的语句是不合理的，但实际上，它是一个有效的 Python 语句，它的求值方式与前面研究的赋值语句类似。具体来说，先对赋值运算符右边的表达式求值，然后把结果存储到赋值运算符左边的变量中。在这个例子中，当语句开始执行时，y 等于 26，因此 y 减去 6 等于 20。然后把 20 存储到 y 中，替换之前存储的 26。后续使用 y 时，将使用新存储的值 20(直到它被另一个赋值语句更改)。

1.2 调用函数

许多程序必须执行一些任务，如从键盘读取输入值、对列表进行排序和计算数字的平方根。Python 提供了执行这些常见任务的函数，以及其他许多函数。我们创建的程序将调用这些函数，这样就不必自己解决这些问题了。

调用函数时,应使用它的名称,后跟括号。许多函数在调用时需要值,例如要排序的名称列表或要计算平方根的数字。这些值称为参数,在调用函数时放在括号内。当一个函数调用有多个参数时,参数之间用逗号分隔。

许多函数计算结果。该结果可以使用赋值语句存储在变量中。变量名出现在赋值操作符的左边,函数调用出现在赋值操作符的右边。例如,下面的赋值语句调用 round() 函数,该函数将一个数字四舍五入为最接近的整数。

```
r = round(q)
```

变量 q(之前必须为其赋值)作为参数传递给 round() 函数。当 round() 函数执行时,它识别最接近 q 的整数并返回它,然后将返回的整数存储在 r 中。

1.2.1 阅读输入

Python 程序可通过调用 input() 函数从键盘读取输入。这个函数使程序停止,并等待用户输入一些内容。当用户按下 Enter 键时,input() 函数就会返回用户输入的字符,然后程序继续执行。输入值通常使用赋值语句存储在变量中,以便以后在程序中使用。例如,下面的语句读取用户输入的值,并将其存储在名为 a 的变量中。

```
a = input()
```

input() 函数总是返回一个字符串,这是计算机科学对字符序列的定义。如果正在读取的值是人名、书名或街道名,那么将该值存储为字符串是合适的。但是,如果值是数字,例如年龄、温度或餐费,那么用户输入的字符串通常会被转换为数字。程序员必须决定转换的结果是整数还是浮点数(可以包括小数点右边的数字)。向整数的转换是通过调用 int() 函数来执行的,而向浮点数的转换是通过调用 float() 函数来执行的。

在从用户读取输入值的同一赋值语句中调用 int() 和 float() 函数是很常见的。例如，下面的语句读取客户名，要购买的物品的数量，以及物品的价格。每个值都用赋值语句存储在自己的变量中。名称存储为字符串，数量存储为整数，价格存储为浮点数。

```
name = input("Enter your name: ")
quantity = int(input("How many items? "))
price = float(input("Cost per item? "))
```

注意，每次调用 input() 函数时，都会向它提供一个参数。这个参数是可选的，是一个提示符，告诉用户输入什么。提示符必须是字符串，用双引号括起来，以便 Python 知道将字符视为字符串，而不是将它们解释为函数或变量的名称。

可以对整数和浮点数进行数学计算。例如，可以使用下面的赋值语句创建另一个变量，来保存物品的总成本。

```
total = quantity * price
```

只有使用前面描述的 int() 和 float() 函数将 quantity 和 price 转换为数字，此语句才会成功执行。试图将这些值相乘而不将它们转换为数字，将导致 Python 程序崩溃。

1.2.2 显示输出

文本输出是使用 print() 函数生成的。可以使用一个参数调用它，该参数是要显示的值。例如，下面的语句打印数字 1，字符串 Hello! 以及当前存储在 x 中的内容。内容可以是整数、浮点数、字符串或还没有讨论过的其他类型的值。每一项都显示在独立的行中。

```
print(1)
print("Hello!")
print(x)
```

通过向 print() 函数提供几个参数，可以通过一个函数调用打印多个值。附加的参数用逗号分隔。例如：

```
print("When x is", x, "the value of y is", y)
```

所有这些值都打印在同一行上。双引号括起的参数是按类型显示的字符串。其他参数是变量。在打印变量时，Python 显示当前存储在其中的值。当打印多项时，每项之间会自动包含一个空格。

函数调用的参数可以是值和变量，如前所述。它们也可以是任意复杂的表达式，包括括号、数学运算符和其他函数调用。考虑下面的语句：

```
print("The product of", x, "and", y, "is", x * y)
```

执行时，计算乘积 x * y，然后与 print() 函数的所有其他参数一起显示出来。

1.2.3 导入其他函数

有些函数，如 input() 和 print() 在许多程序中使用，而另一些函数则没有得到广泛使用。最常用的函数在所有程序中都可用，而其他不太常用的函数则存储在模块中，程序员可以在需要时导入它们。例如，其他数学函数位于 math 模块中。可以在程序开始时包含以下语句来导入该模块：

```
import math
```

math 模块中的函数包括 sqrt()、ceil() 和 sin() 等。从模块导入的函数在调用时，应使用模块名，后跟句号，然后是函数名及其参数。例如，下面的语句调用 math 模块的 sqrt() 函数，计算 y 的平方根(必须在之前进行初始化)，并将结果存储在 z 中。

```
z = math.sqrt(y)
```

其他常用的 Python 模块包括 random、time 和 sys 等。关于这些模块的更多信息可在网上找到。

1.3 注释

注释使程序员有机会解释他们在程序中做什么、如何做或为什么做。这些信息在你离开项目一段时间后又返回项目时非常有用，或者在处理最初由其他人创建的程序时也非常有用。计算机忽略程序中的所有注释。

在 Python 中，注释的开头用#字符表示。注释从#字符一直延续到行尾。注释可占据整行，也可只占一部分，注释出现在 Python 语句的右侧。

Python 文件通常以简单描述程序用途的注释开始。因此，任何人都可以在不仔细检查代码的情况下，快速确定程序的功能。注释代码还可以更容易地确定哪些行执行计算程序结果所需的任务。强烈建议读者在完成本书中的所有练习时，写下详尽的评论。

1.4 值的格式化

有时，数学计算的结果是一个小数点右边有许多位的浮点数。虽然某些程序可能希望显示所有数字，但在其他情况下，值必须四舍五入到特定的小数位数。另一个不相关的程序可能会输出大量需要排列成行的整数。Python 的格式化构造允许完成这些任务，以及许多其他任务。

程序员告诉 Python 如何使用格式说明符格式化值。说明符是描述各种格式细节的字符序列。它使用一个字符来表示应该执行哪种格式。例如，f 表示应该将值格式化为浮点数，d 或 i 表示应该将值格式化为十进制(以 10 为基数)整数，s 表示应该将值格式化为字符串。字符可放在 f、d、i 或 s 之前，以控制其他格式细节。我们只考虑格式化浮点数的问题，使该数在小数点的右边只包括特定数量的数字和格式化值，这样它们占用的字符数最少(允许值整齐地打印出来)。使用格式说明符可以执行其他许多格式化任务，但这些任务超出了本书的范围。

在格式说明符中包含小数点、所需的位数，后面紧跟着 f，就可以将浮点数格式化为包含特定的小数位数。例如，.2f 表示值应该格式化为小数点右边有两位数字的浮点数，而.7f 表示小数点右边应该有 7 位数字。减少小数点右边的位数，就是进行舍入。如果增加数字的位数，则加零。当格式化整数和字符串时，不能指定小数点右侧的位数。

整数、浮点数和字符串都可以进行格式化，使它们至少占据一定的宽度。若生成的输出包含需要对齐的值列，指定最小宽度非常有用。要使用的最小字符数放在 d、i、f 或 s 之前，还可能有小数点前面和小数点后面的位数。例如，8d 表明值应格式化为一个占据至少 8 个字符的十进制整数；而 6.2f 表示，值应格式化为浮点数，至少占用 6 个字符，包括小数点及其右边的两个数字。如有必要，前导空格添加到格式化的值中，以达到最小字符数。

最后，一旦确定了正确的格式字符，就在它们前面加上百分号(%)。格式说明符通常出现在字符串中。可以是字符串中唯一的字符，也可以是更长消息的一部分。完整格式说明符字符串的例子包括 "%8d"、"The amount owing is %.2f" 和 "Hello %s! Welcome aboard!"。

创建格式说明符之后，格式化操作符(用%表示)用于格式化一个值[1]。包含格式说明符的字符串出现在格式操作符的左侧。被格式化的值显示在其右侧。当计算格式化操作符时，将右边的值插入左边的字符串中(在使用指定格式的格式说明符的位置)，来计算操作符的结果。字符串中不属于格式说明符的任何字符都将保留，不需要修改。在格式化操作符左边的字符串中包含多个格式说明符，并在格式化操作符右边的圆括号内用逗号分隔所有要格式化的值，

---

1 Python 提供了几种不同的字符串格式化机制，包括格式化操作符、format 函数和 format 方法、模板字符串，以及最近的 f-字符串。本书中的所有示例和练习都使用格式化操作符，但也可以使用其他技术来实现相同的结果。

可以同时格式化多个值。

字符串格式化通常作为 print 语句的一部分执行。下面代码段中的第一个 print 语句显示变量 x 的值，该变量的小数点右边正好有两位数字。第二个 print 语句先格式化两个值，再把它们显示为较大输出消息的一部分。

```
print("%.2f" % x)
print("%s ate %d cookies!" % (name, numCookies))
```

下表中显示了几个附加的格式化示例。变量 x、y 和 z 之前分别赋值为 12、-2.75 和"Andrew"。

代码段：	`"%d"% x`
结果：	`"12"`
解释：	存储在 x 中的值格式化为十进制(以 10 为基数)整数
代码段：	`"%f"% y`
结果：	`"-2.75"`
解释：	存储在 y 中的值格式化为浮点数
代码段：	`"%d and %f"% (x, y)`
结果：	`"12 and -2.75"`
解释：	存储在 x 中的值格式化为十进制(以 10 为基数)整数，存储在 y 中的值格式化为浮点数。保留字符串中的其他字符，不进行修改
代码段：	`"%.4f" % x`
结果：	`"12.0000"`
解释：	存储在 x 中的值格式化为浮点数，其小数点右边有 4 位数字
代码段：	`"%.1f" % y`
结果：	`"-2.8"`
解释：	存储在 y 中的值格式化为浮点数，其小数点右边有 1 位数字。值在格式化时四舍五入，因为小数点右边的位数减少了
代码段：	`"%10s" % z`
结果：	`"    Andrew"`
解释：	存储在 z 中的值格式化为字符串，因此它至少占用 10 个空格。因为 z 只有 6 个字符长，所以结果中包含了 4 个前导空格
代码段：	`"%4s" % z`
结果：	`"Andrew"`
解释：	存储在 z 中的值格式化为字符串，因此它至少占用 4 个空格。因为 z 比指定的最小长度长，所以结果字符串等于 z

(续表)

代码段:	`"%8i%8i" % (x, y)`
结果:	`"      12      -2"`
解释:	x 和 y 的格式都是十进制(以 10 为基数)整数，占用至少 8 个空格。必要时添加前导空格。当 y(浮点数)格式化为整数时，小数点右边的数字会被截断(而不是四舍五入)

1.5 处理字符串

与数字一样，字符串可以通过操作符操作并传递给函数。通常对字符串执行的操作包括连接两个字符串、计算字符串的长度以及从字符串中提取单个字符。这些通用操作将在本节其余部分描述。有关其他字符串操作的信息可在网上找到。

可以使用+运算符连接字符串。操作符右边的字符串被附加到操作符左边的字符串，以形成新的字符串。例如，下面的程序从用户读取两个字符串，分别是人名和姓氏。然后，它使用字符串连接操作来构造一个新字符串，该字符串是该人的姓，后跟逗号和空格，此后是该人的名。然后显示连接结果。

```
# Read the names from the user
first = input("Enter the first name: ")
last = input("Enter the last name: ")

# Concatenate the strings
both = last + ", " + first

# Display the result
print(both)
```

字符串中的字符数称为字符串的长度。这个值总是一个非负整数，它是通过调用 len()函数来计算的。字符串作为其唯一参数传递给函数，该字符串的长度作为其唯一结果返回。下面的示例通过计算人名的长度来演示 len()函数。

```
# Read the name from the user
```

```
first = input("Enter your first name: ")

# Compute its length
num_chars = len(first)

# Display the result
print("Your first name contains", num_chars, "characters")
```

有时需要访问字符串中的单个字符。例如，为了显示一个人的首字母，可能希望从包含人名、中间名和姓氏的三个字符串中提取第一个字符。

字符串中的每个字符都有一个唯一的整数索引。字符串中第一个字符的索引为0，最后一个字符的索引为字符串长度减去1。通过在包含字符串的变量名之后的方括号内放置其索引，可以访问字符串中的单个字符。下面的程序通过显示人名的首字母来演示这一点。

```
# Read the user's name
first = input("Enter your first name: ")
middle = input("Enter your middle name: ")
last = input("Enter your last name: ")

# Extract the first character from each string and concatenate them
initials = first[0] + middle[0] + last[0]

# Display the initials
print("Your initials are", initials)
```

字符串中的几个连续字符可以通过在方括号中包含两个索引(由冒号分隔)来访问。这称为对字符串进行切片。字符串切片可以有效地访问字符串中的多个字符。

1.6 练习

本章的练习将把之前讨论的概念付诸实践。虽然它们要求完成的任务通常比较小，但是解答这些练习是创建更大程序来解决更有趣问题的重要步骤。

练习 1：邮寄地址

(有解答，9 行)

创建一个程序，显示你的姓名和完整的邮件地址。地址应该按照你住的地方通常使用的格式打印。程序不需要读取来自用户的任何输入。

练习 2：你好

(9 行)

编写一个程序，要求用户输入他或她的名字。程序应该使用用户的姓名来响应一条消息，向用户打招呼。

练习 3：房间的面积

(有解答，13 行)

编写一个程序，要求用户输入房间的宽度和长度。一旦读取这些值，程序就应该计算和显示房间的面积。长度和宽度将作为浮点数输入。在提示和输出消息中包含单位：英尺或米，这取决于你使用哪个单位更舒服。

练习 4：农场的面积

(有解答，15 行)

创建一个程序，读取用户输入的农场的长度和宽度(单位是英尺)。以英亩为单位显示这块地的面积。

> 提示：一英亩有 43 560 平方英尺。

练习 5：瓶子押金

(有解答，15 行)

在许多司法管辖区，会在饮料瓶中加入小额押金，以鼓励人们对其进行回收利用。在一个特定的司法管辖区，装一升或以下饮料的容器有 10 美分的押金，装一升以上饮料的容器有 25 美分的押金。

编写一个程序，从用户那里读取每种饮料瓶的数量。程序应该继续计算和显示将收到的退还饮料瓶的退款。格式化输出，使其包含美元符号，并保留两位小数。

练习 6：征税和小费

(有解答，17 行)

为此练习创建的程序从读取用户在餐厅订购的一顿饭的费用开始，然后程序计算这顿饭的税和小费。在计算应纳税额时使用当地税率。计算小费为餐费的 18%(不含税)。程序的输出应该包括税的金额、小费的金额，以及包括税和小费的总餐费。格式化输出，以便所有值都使用两位小数来显示。

练习 7：前 n 个正整数的和

(有解答，11 行)

编写一个程序，从用户那里读取一个正整数 n，然后显示从 1 到 n 的所有整数的和。前 n 个正整数的和可以用如下公式计算：

$$\text{sum} = \frac{n(n+1)}{2}$$

练习 8：widget 和 gizmo

(15 行)

一个在线零售商销售两种产品：widget 和 gizmo。每个 widget 重 75 克，每个 gizmo 重 112 克。编写一个程序，从用户那里读取 widget 的数量和 gizmo 的数量。然后程序应该计算和显示产品的总重量。

练习 9：复利

(19 行)

假设你刚开了一个新的储蓄账户，每年的利息是 4%。赚的利息在年底支付，然后加到储蓄账户的余额中。编写一个程序，它首先从用户那里读取存入账户的金额。然后，计算并显示储蓄账户在 1 年、2 年和 3 年后的金额。显示每个金额，使其四舍五入到小数点后两位。

练习 10：算术

(有解答，22 行)

创建一个程序，从用户那里读取两个整数 a 和 b。程序应该计算和显示：

- a 和 b 的总和
- 从 a 中减去 b 的差
- a 和 b 的乘积
- a 除以 b 的商
- 当 a 除以 b 时的余数
- $\log_{10} a$ 的结果
- a^b 的结果

提示：math 模块中的 log10() 函数有助于计算列表中的倒数第二项。

练习 11：燃料效率

(13 行)

在美国，车辆的燃油效率通常以每加仑英里数(MPG)表示。在加拿大，燃油效率通常以每百公里升数(L/100 公里)来表示。使用研究技能来决定如何从 MPG 转换到 L/100 公里。然后创建一个程序，从用户那里读取一个美国单位的值，用加拿大单位显示等效的燃油效率。

练习 12：地球上两点之间的距离

(27 行)

地球表面是弯曲的，经度之间的距离随纬度而变化。因此，求地球表面两点之间的距离比简单地使用勾股定理要复杂得多。

设(t_1, g_1)和(t_2, g_2)为地球表面两点的经纬度。这两点在地球表面的距离为(单位是千米)：

$$距离 = 6371.01 \times \arccos(\sin(t_1) \times \sin(t_2) + \cos(t_1) \times \cos(t_2) \times \cos(g_1 - g_2))$$

上述方程中的值 6371.01 不是随机选取的。它是地球的平均半径，以千米为单位。

创建一个程序，允许用户输入地球上两点的经度和纬度(以度为单位)。程序应该显示这两点在地球表面的距离，以千米为单位。

> 提示：Python 的三角函数是以弧度为单位的。因此，在使用前面讨论的公式计算距离之前，需要将用户的输入从角度转换为弧度。math 模块包含一个名为 radians() 的函数，它将角度转换为弧度。

练习 13：找零钱

(有解答，35 行)

考虑在自动结账机上运行的软件。它必须能执行的一项任务是，确定当购物者用现金购买时，需要提供多少零钱。

编写一个程序，首先从用户那里读取一个整数。然后，计算和显示硬币的面额，这些硬币应该用来给购物者提供足够的零钱。零钱应该尽可能少用硬币。假设这台机器装满了 1 分、5 分、10 分、25 分、1 元和 2 元硬币。

一美元硬币于 1987 年在加拿大发行。它称为 loonie，因为硬币的一面有一个潜鸟图像。两美元硬币称为 toonie，是在 9 年后推出的。它的名字来源于数字 2 和 loonie 的组合。

练习 14：身高单位

(有解答，16 行)

许多人以英尺和英寸来计算身高,甚至在一些主要使用公制单位的国家也是如此。编写一个程序,读取用户的英尺数,然后是英寸数。一旦读取了这些值,程序就应该计算并显示相应的厘米数。

> 提示:一英尺是 12 英寸。1 英寸等于 2.54 厘米。

练习 15:距离单位

(20 行)

本练习将创建一个程序,该程序首先读取用户的英尺数据。然后,程序应该以英寸、码和英里显示相同的距离。如果没有记住必要的转换因子,可以上网查找。

练习 16:面积和体积

(15 行)

编写一个程序,首先读取用户的半径 r。程序将计算和显示半径为 r 的圆的面积,以及半径为 r 的球体的体积。在计算中使用 math 模块中的 pi 常数。

> 提示:使用公式 $A=\pi r^2$ 计算圆的面积。使用公式 $V=\frac{4}{3}\pi r^3$ 计算球的体积。

练习 17:比热容

(有解答,23 行)

使 1 克材料的温度提高 1 摄氏度所需的能量是材料的比热容 C。使 m 克材料的温度提高 ΔT 摄氏度所需的总能量 Q 使用如下公式计算:

$$Q = mC\Delta T$$

编写一个程序,从用户那里读取水的质量和温度变化。程序应该显示为实现所需的温度变化而必须添加或删除的总能量。

提示：水的比热容是 $4.186\dfrac{\text{J}}{\text{g}\cdot{}^\circ\text{C}}$。因为水的密度是 1.0 克/毫升，所以在这个练习中可互换使用克和毫升。

扩展程序，以计算加热水的成本。电流表通常使用的单位是千瓦时，而不是焦耳。在这个练习中，假设电的成本是 8.9 美分/千瓦时。用程序来计算煮一杯咖啡所需要的加热水的成本。

提示：需要查找焦耳和千瓦时的转换因子，来完成这个练习的最后一部分。

练习 18：圆柱体的体积

(15 行)

圆柱体的体积可以用圆底面积乘以高来计算。编写一个程序，从用户那里读取圆柱体的半径和高，并计算其体积。显示四舍五入到小数点后一位的结果。

练习 19：自由落体

(有解答，15 行)

创建一个程序来确定物体落地时的速度。用户输入物体下落的高度，单位为米，因为物体下落的初始速度为 0 米/秒。假设重力加速度是 9.8 米/秒^2。当初始速度 v_i、加速度 a 和距离 d 已知时，可用公式 $v_f = \sqrt{v_i^2 + 2ad}$ 来计算最终速度 v_f。

练习 20：理想气体定律

(19 行)

理想气体定律是气体随压力、体积和温度变化的数学近似。它通常表述为：

$$PV = nRT$$

P 是压强，单位是 Pa，V 是体积，单位是 L，n 是气体量，单位是 mol，R 是理想气体常数，等于 $8.314 \dfrac{\text{J}}{\text{mol} \cdot \text{K}}$，$T$ 是温度，用开氏度表示。

编写一个程序，当用户提供了压力、体积和温度时，计算气体量，单位是 mol。测试程序：确定 SCUBA 水箱中气体的摩尔数。典型的 SCUBA 水箱能在 2000 万 Pa(约 3000PSI)的压力下装 12L 气体。室温大约是 20 摄氏度或 68 华氏度。

> 提示：给温度加上 273.15，摄氏度就转换成开氏温度。要把华氏温度转换成开氏温度，要减去 32，再乘以 $\dfrac{5}{9}$，然后加上 273.15。

练习 21：三角形的面积

(13 行)

三角形的面积可用以下公式计算，其中 b 是三角形的底，h 是三角形的高。

$$\text{area} = \dfrac{b \times h}{2}$$

编写一个程序，允许用户输入 b 和 h 的值。然后该程序应计算和显示底为 b、高为 h 的三角形的面积。

练习 22：三角形的面积

(16 行)

在前面的练习中，创建了一个程序，该程序在已知三角形的底和高时计算三角形的面积。当已知三角形的三边长度时，也可以计算三角形的面积。设 s_1、s_2、s_3 为边的长度。设 $s = (s_1 + s_2 + s_3)/2$。

然后用以下公式计算三角形的面积：

$$\text{area} = \sqrt{s \times (s-s_1) \times (s-s_2) \times (s-s_3)}$$

开发一个程序,从用户那里读取三角形的边长,并显示其面积。

练习 23：正多边形的面积

(有解答，14 行)

如果一个多边形的所有边长都相同,并且所有相邻边之间的夹角都相等,那么这个多边形就是正多边形。正多边形的面积可以用以下公式计算,其中 s 为边长, n 为边的个数：

$$\text{area} = \frac{n \times s^2}{4 \times \left(\dfrac{\pi}{n}\right)}$$

编写一个程序,从用户那里读取 s 和 n,然后显示由这些值构成的正多边形的面积。

练习 24：计算持续时间

(22 行)

创建一个程序,将用户的持续时间读取为天数、小时、分钟和秒。计算并显示此持续时间所表示的总秒数。

练习 25：时间单位

(有解答，24 行)

本练习将逆转练习 24 的过程。开发一个程序,首先读取用户输入的秒数。然后,程序应该以 D：HH：MM：SS 的形式显示等量的时间,其中 D、HH、MM 和 SS 分别代表天数、小时、分钟和秒。小时、分钟和秒都应该格式化,使它们刚好显示为两位数字。使用研究技能确定,需要在格式说明符中包含哪些额外字符,以便在数字格式化为特定宽度时，使用前导零而不是前导空格。

练习 26：当前时间

(10 行)

Python 的 time 模块包括几个与时间相关的函数。其中之一是 asctime()函数，它从计算机的内部时钟读取当前时间，并以人类可读的格式返回。使用此函数编写一个程序，显示当前时间和日期。程序不需要用户输入。

练习 27：复活节在什么时候？

(33 行)

复活节是春分后第一个月圆后的那个星期天。因为复活节的日期包含了月亮的成分，所以在公历中没有固定的日期。相反，它可以在 3 月 22 日和 4 月 25 日之间的任何日期。可以使用 Anonymous Gregorian Computus 算法计算给定年份的复活节的月份和日期，如下所示。

设 a 等于年份除以 19 的余数

设 b 等于年份除以 100 的取整商

设 c 等于年份除以 100 的余数

设 d 等于 b 除以 4 的取整商

设 e 等于 b 除以 4 的余数

设 f 等于 $\dfrac{b+8}{25}$ 的取整商

设 g 等于 $\dfrac{b-f+1}{3}$ 的取整商

设 h 等于 $19a+b-d-g+15$ 除以 30 的余数

设 i 等于 c 除以 4 的取整商

设 k 等于 c 除以 4 的余数

设 l 等于 $32+2e+2i-h-k$ 除以 7 的余数

设 m 等于 $\dfrac{a+11h+22l}{451}$ 的取整商

设 month 等于 $\dfrac{h+l-7m+114}{31}$ 的取整商

设 day 等于 $\dfrac{h+l-7m+114}{31}$ 的余数

编写一个程序，实现 Anonymous Gregorian Computus 算法，来计算复活节的日期。程序应该读取用户输入的年份，然后显示一个适当消息，其中包括当年的复活节日期。

练习 28：身体质量指数

(14 行)

编写一个程序来计算一个人的身体质量指数(BMI)。程序应首先读取用户的身高和体重，然后使用以下两个公式之一计算 BMI，之后显示它。如果用英寸表示身高，用磅表示体重，那么体重指数的计算方法如下。

$$BMI = \dfrac{weight}{height \times height} \times 703$$

如果用米表示身高，用千克表示体重，那么体重指数就可用下面这个稍微简单一点的公式来计算。

$$BMI = \dfrac{weight}{height \times height}$$

练习 29：风寒

(有解答，22 行)

在寒冷的天气里有风时，空气感觉比实际更冷，因为空气的流动增加了温暖物体(比如人)的冷却速度，这种效应称为风寒。

2001 年，加拿大、英国和美国采用了如下公式来计算风寒指数。式中 T_a 为气温，单位为摄氏度；V 为风速，单位为千米/小时。类似的公式可以用不同的恒定值来表示华氏温度和风速(英里/小时)。

$$WCI = 13.12 + 0.6215T_a - 11.37V^{0.16} + 0.3965T_a V^{0.16}$$

编写一个程序，首先读取用户输入的气温和风速。一旦读取了这些值，程序就应该显示风寒指数(四舍五入到最接近的整数)。

> **提示**：风寒指数只适用于气温低于或等于 10 摄氏度，风速超过 4.8 千米/小时的情况。

练习 30：摄氏温度转换成华氏温度和开尔文温度

(17 行)

编写一个程序，首先读取用户输入的温度(以摄氏度为单位)。然后程序应该以华氏度和开尔文度显示等效温度。转换不同温度单位所需的计算公式可在因特网上找到。

练习 31：压力的单位

(20 行)

这个练习将创建一个程序，来读取用户输入的千帕斯卡压力。一旦读取了压力，程序应该报告等效的压力(以磅/平方英寸、毫米汞柱和大气压为单位)。利用研究技能来确定这些单位之间的转换系数。

练习 32：整数中各个数字的和

(18 行)

开发一个程序，从用户那里读取一个四位数的整数并显示其位数的和。例如，如果用户输入 3141，那么程序应该显示 3 + 1 + 4 + 1 = 9。

练习 33：对 3 个整数排序

(有解答，19 行)

创建一个程序，读取用户输入的三个整数，并按顺序(从小到大)显示它们。使用 min() 和 max() 函数查找最小值和最大值。通过计算三个值的和，然后减去最小值和最大值，可以找到中间值。

练习 34：旧面包

(有解答，19 行)

一家面包店以 3.49 美元/条的价格出售面包。旧面包打六折。编写一个程序，首先读取用户购买的面包数量。然后显示面包的正常价格，已存放一天的折扣价，以及总价格。这些金额中的每一个都应该显示在单独一行中，并带有适当标签。所有值都应该使用两位小数来显示，当用户输入合理值时，所有数字中的小数点应该对齐。

第2章 决 策

第1章使用的程序是严格按顺序排列的。每个程序的语句都是按顺序执行的,从程序的开始执行到结束,没有中断。虽然程序中的每条语句按顺序执行可以用来解答一些小练习,但它不足以解决大多数有趣的问题。

决策结构允许程序包含的语句在程序运行时可能执行,也可能不执行。程序仍然从顶部开始向底部执行,但是可能跳过程序中的一些语句。这允许程序对不同的输入值执行不同的任务,并极大地增加了 Python 程序可以解决的问题范围。

2.1 if 语句

Python 程序使用 if 语句进行决策。if 语句包含一个条件和一个或多个构成 if 语句主体的语句。当执行 if 语句时,将对其条件进行评估,以确定其主体中的语句是否执行。如果条件的值为 True,则执行 if 语句的主体,然后执行程序中的其余语句。如果 if 语句的条件计算结果为 False，则跳过 if 语句的主体，并继续执行 if 语句主体之后的第一行。

if 语句的条件可以是任意复杂的表达式，其计算结果为 True 或 False。这种表达式被称为布尔表达式，以形式逻辑的先驱乔治·布尔(1815—1864)的名字命名。if 语句的条件通常包含一个关系运算符，用于比较两个值、变量或复杂表达式。下面列出 Python 的关系运算符。

关系运算符	含义
<	小于
<=	小于或等于
>	大于
>=	大于或等于
==	等于
!=	不等于

if 语句的主体由一条或多条语句组成，这些语句的缩进必须大于 if 关键字。在缩进量等于(或小于)if 关键字的下一行语句结束。可以选择在缩进 if 语句主体时使用多少空格。本书中介绍的所有程序都缩进两个空格，但如果愿意，可以使用一个空格，也可以使用多个空格[1]。

下面的程序读取用户输入的一个数字，使用两条 if 语句将描述该数字的消息存储到 result 变量中，然后显示该消息。每条 if 语句的条件都使用一个关系运算符来确定其缩进的主体是否会执行。冒号紧跟在每个条件后面，将 if 语句的条件与其主体分隔开。

```
# Read a number from the user
num = float(input("Enter a number: "))

# Store the appropriate message in result
```

---

1 大多数程序员选择给 if 语句的主体缩进相同数量的空格，但是 Python 不需要这种一致性。

```
if num == 0:
   result = "The number was zero"
if num != 0:
   result = "The number was not zero"

# Display the message
print(result)
```

2.2 if - else 语句

前一个示例在用户输入的数字为 0 时，将一条消息存储到 result 中，在输入的数字不为 0 时，将另一条消息存储到 result 中。if 语句上的条件是这样构造的：两个 if 语句体中只有一个会执行。无法使两个主体都执行，也无法使两个主体都不执行(这些条件被称为互斥)。

if-else 语句由带条件和主体的 if 部分和带主体(但不带条件)的 else 部分组成。当语句执行时，它的条件被求值。如果条件的值为 True，则执行 if 部分的主体，跳过 else 部分的主体。当条件的计算结果为 False 时，将跳过 if 部分的主体，并执行 else 部分的主体。不可能同时执行两个主体，也不可能同时跳过两个主体。因此，当一条 if 语句紧跟着另一条 if 语句时，可以使用 if-else 语句而不是两条 if 语句，而且 if 语句上的条件是互斥的。最好使用 if-else 语句,因为只需要编写一个条件,在程序执行时只需要计算一个条件，如果将来发现错误，只需要纠正一个条件。下面的程序报告值是否为零，重写后使用 if-else 语句。

```
# Read a number from the user
num = float(input("Enter a number: "))

# Store the appropriate message in result
if num == 0:
   result = "The number was zero"
else:
   result = "The number was not zero"
```

```
# Display the message
print(result)
```

当用户输入的数字为 0 时，if-else 语句上的条件计算为 True，因此执行语句的 if 部分，并将适当的消息存储到 result 中。然后，跳过语句的 else 部分主体。当数字非零时，if-else 语句上的条件将计算为 False，因此将跳过 if 语句的主体部分。由于跳过了 if 部分的主体，因此执行 else 部分的主体，从而将另一个消息存储到 result 中。这两种情况下，Python 都会继续运行程序的其余部分，以显示消息。

2.3　if-elif-else 语句

if-elif-else 语句用来执行几个可选方案中的一个。该语句以 if 部分开头，然后是一个或多个 elif 部分，再后是 else 部分。所有这些部分都必须包括一个缩进的主体。每个 if 和 elif 部分还必须包含一个条件，该条件的计算结果为 True 或 False。

当执行 if-elif-else 语句时，首先对 if 部分的条件求值。如果计算结果为 True，则执行 if 部分的主体，并跳过所有 elif 和 else 部分。但是，如果 if 部分的条件计算为 False，则跳过它的主体，然后 Python 继续对第一个 elif 部分计算条件。如果此条件的值为 True，则执行第一个 elif 部分的主体，并跳过所有其余的条件和主体。否则，Python 将继续依次计算每个 elif 部分上的条件。这将一直进行下去，直到找到一个计算结果为 True 的条件。然后执行与该条件关联的主体，并跳过其余的 elif 和 else 部分。如果 Python 到达语句的 else 部分(因为 if 和 elif 部分的所有条件的计算结果为 False)，那么它将执行 else 部分的主体。

扩展前面的示例，以便为正数显示一条消息，为负数显示另一条消息，如果数字为 0，则显示另一条不同的消息。虽然可以使用

if 和/或 if-else 语句的组合来解决这个问题，但是这个问题非常适合 if-elif-else 语句，因为必须执行三个选项中的一个。

```
# Read a number from the user
num = float(input("Enter a number: "))

# Store the appropriate message in result
if num > 0:
    result = "That's a positive number"
elif num < 0:
    result = "That's a negative number"
else:
    result = "That's zero"

# Display the message
print(result)
```

当用户输入一个正数时，语句的 if 部分的条件计算为 True，因此执行 if 部分的主体。一旦执行了 if 部分的主体，程序将执行最后一行的 print 语句，跳过了 elif 部分和 else 部分的主体，而没有计算语句的 elif 部分的条件。

当用户输入一个负数时，语句 if 部分的条件计算为 False。Python 跳过 if 部分的主体，继续计算语句的 elif 部分的条件。该条件的值为 True，因此执行 elif 部分的主体。然后跳过 else 部分，程序继续执行 print 语句。

最后，当用户输入 0 时，语句 if 部分的条件将计算为 False，因此跳过 if 部分的主体，Python 将继续计算 elif 部分的条件。它的条件也计算为 False，因此 Python 继续执行 else 部分的主体。然后执行最后的 print 语句。

在任意数量的选项中，只有一个选项是由 if-elif-else 语句执行的。该语句以 if 部分开头，然后根据需要使用尽可能多的 elif 部分。else 部分总是最后出现，它的主体只在 if 和 elif 部分上的所有条件都为 False 时执行。

2.4 if-elif 语句

出现在 if-elif-else 语句末尾的 else 是可选的。当 else 出现时，语句会从几个选项中选择一个。省略 else 至多选择几个选项中的一个。当使用 if-elif 语句时，当所有条件的值都为 False 时，不执行任何主体。无论是否执行其中一个主体，程序都会继续执行最后一个 elif 部分的主体之后的第一个语句。

2.5 嵌套的 if 语句

任何 if 部分的主体、任何类型 if 语句的 elif 部分或 else 部分都可以包含几乎任何 Python 语句，包括另一个 if、if-else、if-elif 或 if-elif-else 语句。当一个 if 语句(任意类型)出现在另一个 if 语句(任意类型)的主体中时，就说 if 语句是嵌套的。下面的程序包含一个嵌套的 if 语句。

```
# Read a number from the user
num = float(input("Enter a number: "))

# Store the appropriate message in result
if num > 0:

    # Determine what adjective should be used to describe the number
    adjective = " "
    if num >= 1000000:
        adjective = " really big "
    elif num >= 1000:
        adjective = " big "

    # Store the message for positive numbers including the appropriate adjective
    result = "That's a" + adjective + "positive number"
elif num < 0:
    result = "That's a negative number"
else:
    result = "That's zero"
```

```
# Display the message
print(result)
```

这个程序首先从用户那里读取一个数字。如果用户输入的数字大于 0，则执行外部 if 语句的主体。它首先将一个包含一个空格的字符串赋给 adjective。然后执行嵌套在 if-elif-else 语句内部的 if-elif 语句。如果输入的数字至少为 1 000 000，则内部语句将 adjective 更新为 really big；如果输入的数字在 1 000 到 999 999 之间，则将 adjective 更新为 big。外部 if 部分的主体中的最后一行将完整消息存储在 result 中，然后由于执行了外部 if 部分的主体，因此将跳过外部 elif 部分的主体和外部 else 部分的主体。最后，程序通过执行 print 语句完成。

现在考虑一下，如果用户输入的数字小于或等于 0 会发生什么。当发生这种情况时，将跳过外部 if 语句的主体，并执行外部 elif 部分的主体或 else 部分的主体。这两种情况都会在 result 中存储适当的消息，然后在程序结束时继续执行 print 语句。

2.6 布尔逻辑

布尔表达式是计算结果为 True 或 False 的表达式。表达式可以包含各种各样的元素，例如布尔值 True 和 False、包含布尔值的变量、关系运算符以及对返回布尔结果的函数的调用。布尔表达式还可以包括组合和操作布尔值的布尔运算符。Python 包含三个布尔运算符，即 not、and 和 or。

not 操作符反转布尔表达式的真值。如果表达式 x 出现在 not 操作符的右边，它的值为 True，于是 not x 的求值结果为 False。如果 x 的值为 False，那么 not x 的值为 True。

任何布尔表达式的行为都可以用真值表来描述。真值表对于布尔表达式中的每个不同的变量都有一列，对于表达式本身也有一列。真值表中的每一行表示表达式中变量的真值和假值的一个组合。具有 n 个不同变量的表达式的真值表有 2^n 行，每一行都显示由不同值组合的表达式计算的结果。not 操作符应用于单个变量 x，其真值表有 $2^1 = 2$ 行，如下所示。

x	not x
False	True
True	False

and 和 or 操作符组合两个布尔值来计算布尔结果。如果 x 为真 (True)，y 也为真，则布尔表达式 x and y 的值为真。如果 x 为假 (False)，或者 y 为假，或者 x 和 y 都为假，那么 x and y 的值为假。and 运算符的真值表如下所示。它有 $2^2 = 4$ 行，因为 and 运算符应用于两个变量。

x	y	x and y
False	False	False
False	True	False
True	False	False
True	True	True

如果 x 为真，或者 y 为真，或者 x 和 y 都为真，那么布尔表达式 x or y 的值为真。它只在 x 和 y 都为假的情况下计算为假。or 操作符的真值表如下所示：

x	y	x or y
False	False	False
False	True	True
True	False	True
True	True	True

下面的 Python 程序使用 or 操作符来确定用户输入的值是不是

前 5 个素数之一。在构造复杂条件时，and 和 not 操作符可用类似的方式使用。

```
# Read an integer from the user
x = int(input("Enter an integer: "))
```

```
# Determine if it is one of the first 5 primes and report the result
if x == 2 or x == 3 or x == 5 or x == 7 or x == 11:
  print("That's one of the first 5 primes.")
else:
  print("That is not one of the first 5 primes.")
```

2.7 练习

下面的练习应该使用 if、if-else、if-elif 和 if-elif-else 语句以及第 1 章介绍的概念来完成。在一些解中，将 if 语句嵌套到另一个 if 语句的主体中也很有帮助。

练习 35：偶数还是奇数？

(有解答，13 行)

编写一个程序，读取用户输入的一个整数。然后，程序应该显示一条消息，指示整数是偶数还是奇数。

练习 36：狗年

(22 行)

人们常说，人的一年相当于狗的 7 年。然而，这个简单的转换并没有考虑到，狗在大约两年后就成年了。因此，有些人认为，最好把前两年的每一年都算作 10.5 个狗年，然后把每一个额外的人类年算作 4 个狗年。

编写一个程序，实现前一段描述的从人类年到狗年的转换。确保程序在少于两年的人类年的转换和超过两年的人类年的转换中正常工作。如果用户输入负数，程序应该显示适当的错误消息。

练习 37：元音或辅音

(有解答，16 行)

本练习将创建一个从用户那里读取字母的程序。如果用户输入 a、e、i、o 或 u，那么程序应该显示一条消息，指示输入的字母是元音。如果用户输入 y，程序应该显示一条消息，指示有时 y 是元音，有时 y 是辅音。否则，程序应该显示一条消息，指示字母是辅音字母。

练习 38：形状的命名

(有解答，31 行)

编写一个程序，从形状的边数来确定形状的名称。读取用户输入的边数，然后将适当的名称作为有意义消息的一部分报告。程序应该支持 3～10 条边的任意形状。如果输入了此范围之外的多条边，则程序应显示适当的错误消息。

练习 39：月名到天数

(有解答，18 行)

一个月的长度从 28 天到 31 天不等。本练习将创建一个程序，从用户那里读取一个月的名称作为字符串。然后程序应该显示该月的天数。给 2 月份显示 "28 或 29 天"，这样闰年就有了。

练习 40：音量

(30 行)

下表列出几种常见噪音的分贝级。

噪音	分贝水平
手提钻	130 分贝
割草机	106 分贝
闹钟	70 分贝
安静房间	40 分贝

第 2 章 决 策

编写一个程序，从用户那里读取一个声音的分贝级别。如果用户输入的分贝级别与表中的噪声之一匹配，则程序应该显示只包含该噪声的消息。如果用户输入的若干分贝在列出的噪声之间，则程序应显示一条消息，指示该值位于哪两个噪声之间。确保程序还生成了合理的输出，其值小于表中最安静的噪声，而大于表中最嘈杂的噪声。

练习 41：三角形的分类

(有解答，21 行)

三角形可以根据其边长分为等边三角形、等腰三角形或不等边三角形。等边三角形的三条边长度相等。等腰三角形的两条边长度相等，而第三条边的长度不相等。如果所有边的长度都不相等，那么这个三角形就是不等边三角形。

编写一个程序，从用户那里读取三角形的三条边长。然后显示一条声明三角形类型的消息。

练习 42：音符的频率

(有解答，39 行)

下表列出一个八度音阶及其频率。

注意	频率(Hz)
C4	261.63
D4	293.66
E4	329.63
F4	349.23
G4	392.00
A4	440.00
B4	493.88

首先编写一个程序，从用户那里读取音符的名称，并显示音符

的频率。程序应该支持前面列出的所有音符。

一旦程序能够正确地处理前面列出的音符，就应该添加对从 C0 到 C8 的所有音符的支持。虽然这可以通过在 if 语句中添加许多额外的情况来实现，但是这样的解决方案非常麻烦、不优雅，对于本练习的目的来说是不可接受的。相反，应该利用相邻八度音阶之间的关系。在八度音阶中，任何一个音符的频率都是后一个八度音阶对应音符的一半。通过使用这个关系，应该能够添加对附加音符的支持，而不必向 if 语句添加额外的情况。

> **提示**：在完成此练习时，可访问用户单独输入的音符中的字符。首先将字母与八度分开，然后使用上表中的数据计算第四个八度音阶中那个字母的频率。一旦有了这个频率，就应该除以 2^{4-x}，其中 x 是用户输入的八度数。这将使正确次数的频率减半或加倍。

练习 43：音符频率的逆转换

(有解答，42 行)

在前一个问题中，将一个音符的名称转换为它的频率。在这个问题中，你要写一个程序来逆转这个过程。首先读取用户输入的频率。如果频率在上题表中列出的值的 1Hz 内，那么报告相应的音符名称。否则报告的频率不对应已知的音符。在这个练习中，只需要考虑表中列出的音符。不需要考虑其他八度音阶的音符。

练习 44：钞票上的脸

(31 行)

国家的前任领导人或其他非政治意义的重要人物形象出现在钞票上是很常见的。在美国，出现在纸币上的个人如下所示。

个人	钞票面额
乔治·华盛顿	1 美元
托马斯·杰斐逊	2 美元

第2章 决 策

(续表)

个人	钞票面额
亚伯拉罕·林肯	5 美元
亚历山大·汉密尔顿	10 美元
安德鲁·杰克逊	20 美元
尤利西斯·辛普森·格兰特	50 美元
本杰明·富兰克林	100 美元

编写一个程序，从读取用户的钞票面额开始。然后，程序应该显示出现在对应钞票面额上的人名。如果不存在这样的钞票，则应该显示适当的错误消息。

虽然在美国流通的两美元纸币很少见，但它们是可以像其他货币一样使用的法定货币。美国还发行面值为 500 美元、1000 美元、5000 美元、10 000 美元的纸币，供公众使用。然而，高面额纸币自 1945 年以来就没有发行过，并于 1969 年正式停用。因此，这个练习不会考虑它们。

练习 45：假日的日期

(18 行)

下面列出加拿大的三个全国性假日，每年都在同一天。

假日	日期
元旦	1 月 1 日
加拿大国庆日	7 月 1 日
圣诞节	12 月 25 日

编写一个程序，读取用户输入的月份和日期。如果月份和日期与前面列出的假日之一匹配，那么程序应该显示假日的名称。否则，程序应指示输入的月份和日期不对应于固定日期的假日。

加拿大还有另外两个全国性假日：耶稣受难日和劳动节，它们的日期每年都不同。还有许多省和地区的节日,有些有固定的日期,有些则有可变的日期。这个练习不会考虑这些额外的假日。

练习 46：那个正方形是什么颜色？

(22 行)

棋盘上的位置由字母和数字来确定。字母表示列,数字表示行,如下图所示。

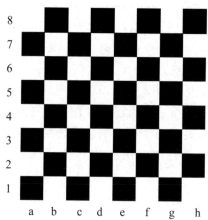

编写一个程序,读取用户输入的位置。使用 if 语句确定列的开头是黑色方块还是白色方块。然后使用模运算来报告该行中正方形的颜色。例如,如果用户输入 a1,那么程序应该报告正方形是黑色的。如果用户输入 d5,那么程序应该报告正方形是白色的。程序可能会假定用户总是输入一个有效位置。它不需要执行任何错误检查。

练习 47：季节的确定日期

(有解答,43 行)

一年分为四季：春、夏、秋、冬。虽然由于年历的构造方式不同,季节变化的确切日期每年略有不同,但本练习使用以下日期。

第 2 章 决 策

季节	第一天
春季	3 月 20 日
夏季	6 月 21 日
秋季	9 月 22 日
冬季	12 月 21 日

创建一个程序,读取用户输入的月份和日期。用户以字符串形式输入月份的名称,然后以整数形式输入月份中的日期。此后,程序应该显示与输入日期相关的季节。

练习 48:出生日期到星座的对应

(47 行)

报纸上经常报道的占星术是利用一个人出生时的太阳位置来预测未来。这种占星术将一年分为十二星座,如下表所示。

星座	日期范围
摩羯座	12 月 22 日至 1 月 19 日
水瓶座	1 月 20 日至 2 月 18 日
双鱼座	2 月 19 日至 3 月 20 日
白羊座	3 月 21 日至 4 月 19 日
金牛座	4 月 20 日至 5 月 20 日
双子座	5 月 21 日至 6 月 20 日
巨蟹座	6 月 21 日至 7 月 22 日
狮子座	7 月 23 日至 8 月 22 日
处女座	8 月 23 日到 9 月 22 日
天秤座	9 月 23 日至 10 月 22 日
天蝎座	10 月 23 日至 11 月 21 日
射手座	11 月 22 日至 12 月 21 日

编写一个程序,要求用户输入其出生日期和月份。然后,程序

应该报告用户的星座，作为适当的输出消息的一部分。

练习 49：十二生肖

(有解答，40 行)

中国的十二生肖以 12 年为一个周期。下表显示了一个 12 年的周期。2012 年是龙年，1999 年是兔年。

年	生肖
2000 年	龙
2001 年	蛇
2002 年	马
2003 年	羊
2004 年	猴
2005 年	鸡
2006 年	狗
2007 年	猪
2008 年	鼠
2009 年	牛
2010 年	虎
2011 年	兔

编写一个程序，读取用户输入的年份，并显示与那年相关的生肖。程序应该在大于或等于 0 的任何年份都能正常工作，而不仅是表中列出的年份。

练习 50：里氏震级

(30 行)

下表列出里氏震级范围及其描述符。

震级	描述符
小于 2.0	超微
2.0 到 3.0(不含)	甚微
3.0 到 4.0(不含)	微小
4 到 5.0(不含)	弱
5.0 至 6.0(不含)	中
6.0 至 7.0(不含)	强
7.0 到 8.0(不含)	甚强
9.0 到 10(不含)	极强
10 或更多	超强

编写一个程序，从用户那里读取震级，并显示适当的描述符，作为有意义的消息的一部分。例如，如果用户输入 5.5，程序应该表明，5.5 级被认为是中度地震。

练习 51：二次函数的根

(24 行)

一元二次函数的形式是 $f(x) = ax^2 + bx + c$，其中 a、b 和 c 是常数，a 非零。通过找到满足二次方程 $ax^2 + bx + c = 0$ 的 x 的值，可以确定它的根。这些值可以用二次公式计算，如下所示。二次函数可以有 0、1 或 2 个实根。

$$\text{root} = \frac{-b \pm \sqrt{b^2 - 4ac}}{2a}$$

根号下的部分称为判别式。如果判别式是负的，那么二次方程就没有实根。如果判别式是 0，那么方程有一个实根。否则，方程有两个实根，当计算分子时，表达式必须求两次值，一次用加号，一次用减号。

写一个计算二次函数实根的程序。程序应该先提示用户输入 a、b 和 c 的值。然后它应该显示一条消息，指示实根的数量，以及实根的值(如果有)。

练习 52：字母等级到绩点

(有解答，52 行)

在一所特定的大学，字母成绩采用下方式映射到绩点。

字母	绩点
A+	4.0
A	4.0
A-	3.7
B+	3.3
B	3.0
B-	2.7
C+	2.3
C	2.0
C-	1.7
D+	1.3
D	1.0
F	0

编写一个程序，首先读取用户输入的字母等级。然后程序应该计算和显示相等数量的绩点。确保程序在用户输入无效字母等级时生成适当的错误消息。

练习 53：从绩点到字母等级

(47 行)

前面的练习创建了一个程序，该程序将字母等级转换为相等数量的绩点。本练习将创建一个程序来逆转这个过程，并将用户输入的绩点值转换为字母级别。确保程序处理介于字母等级之间的绩点值。这些绩点值应该四舍五入到最接近的字母等级。如果用户输入的值是 4.0 或更高，程序应该报告 A+。

练习 54：评估员工业绩

(有解答，30 行)

在一家特定的公司，员工业绩在每年年底都会被评估。评分表从 0.0 开始，较高数值表示更好的业绩，并导致更大的加薪。奖励给员工的值是 0.0、0.4 或 0.6 或更多。永远不会使用 0.0 和 0.4 之间以及 0.4 和 0.6 之间的值。与每个评级相关的含义如下表所示。员工加薪的金额是 2400.00 美元乘以他们的评级。

评级	含义
0.0	不可接受的业绩
0.4	可接受的业绩
0.6 或以上	非常好的业绩

编写一个程序，从用户那里读取评级，并指出该评级的业绩是不可接受的、可接受的还是非常好的。也应该报告员工的加薪金额。如果输入了无效的评级，程序应该显示适当的错误消息。

练习 55：可见光的波长

(38 行)

可见光的波长范围是 380～750nm。光谱是连续的，通常分为 6 种颜色，如下图所示。

颜色	波长(nm)
紫色	380～450(不含)
蓝色	450～495(不含)
绿色	495～570(不含)
黄色	570～590(不含)
橙色	590～620(不含)
红色	620～750

编写一个程序,读取用户输入的波长并报告其颜色。如果用户输入的波长在可见光谱之外,则显示适当的错误消息。

练习 56:按频率来命名

(31 行)

电磁辐射按其频率可分为 7 类,如下表所示。

名称	频率范围(Hz)
无线电波	小于 3×10^9
微波	$3 \times 10^9 \sim 3 \times 10^{12}$(不含)
红外光	$3 \times 10^{12} \sim 4.3 \times 10^{14}$(不含)
可见光	$4.3 \times 10^{14} \sim 7.5 \times 10^{14}$(不含)
紫外线	$7.5 \times 10^{14} \sim 3 \times 10^{17}$(不含)
X 射线	$3 \times 10^{17} \sim 3 \times 10^{19}$(不含)
伽马射线	3×10^{19} 或以上

编写一个程序,读取来自用户的辐射频率,并显示辐射的名称,作为适当消息的一部分。

练习 57:手机账单

(44 行)

一个特别的手机套餐包括 50 分钟的通话时间和 50 条短信,每月 15 美元。每增加一分钟的通话时间,费用为 0.25 美元;每增加一条短信,费用为 0.15 美元。所有手机话费都包括额外的 $0.44 费用,用于支持 911 呼叫中心,整个账单(包括 911 费用)需要缴纳 5% 的销售税。

编写一个程序,读取用户在一个月内使用的分钟数和文本消息。显示基本收费、额外分钟收费(如有)、额外的短信费用(如有)、911 费用、税金和账单总额。如果用户在这些类别中产生了费用,则只显示额外的分钟和文本消息费用。确保所有费用都用两位小数

第2章 决　策

表示。

练习 58：这是闰年吗？

(有解答，22 行)

大多数年份有 365 天。然而，地球绕太阳公转所需的时间实际上比这稍长一些。因此，在某些年份中会增加 2 月 29 日这一天，以弥补这一差异。这样的年份被称为闰年。决定一年是否为闰年的规则如下：

- 任何能被 400 整除的年份都是闰年。
- 在剩下的年份中，任何能被 100 整除的年份都不是闰年。
- 在剩下的年份中，任何能被 4 整除的年份都是闰年。
- 其他年份都不是闰年。

编写一个程序，读取用户输入的年份，并显示一条消息，表明它是不是闰年。

练习 59：第二天

(50 行)

编写一个程序，从用户那里读取数据，并计算它的直接后续数据。例如，如果用户输入代表 2019-11-18 的值，那么程序应该显示一条消息，指示在 2019-11-18 之后的那一天是 2019-11-19。如果用户输入的值代表 2019-11-30，则程序应指示第二天是 2019-12-01。如果用户输入代表 2019-12-31 的值，则程序应指示第二天是 2020-01-01。日期将以数字形式输入，有三个单独的输入语句；分别输入年份、月份和日期。确保程序在闰年工作正常。

练习 60：1 月 1 日是星期几？

(32 行)

下面的公式可用来确定某一年的 1 月 1 日是星期几。

day_of_the_week = (year+ floor((year−1)/ 4)−floor((year−1)/ 100)+
    floor((year−1)/ 400))% 7

这个公式计算的结果是一个表示星期几的整数。星期日用 0 表示。接下来一周里剩下的六天依次表示从星期一到星期六，用 1~6 表示。

使用上面的公式编写一个程序，该程序读取用户输入的年份，并报告该年份的 1 月 1 日是星期几。程序的输出应该包括星期几的全名，而不只是公式返回的整数。

练习 61：车牌有效吗？

(有解答，28 行)

在一个特定的司法管辖区，旧的车牌由三个大写字母和三个数字组成。当使用所有遵循该模式的牌照后，格式更改为四个数字后面跟着三个大写字母。

编写一个程序，首先读取用户输入的字符串。然后，程序应该显示一条消息，指示这些字符是对旧样式的车牌有效还是对新样式的车牌有效。如果用户输入的字符串对两种样式的车牌都是无效的，程序就应该显示一个适当的消息。

练习 62：轮盘赌

(有解答，45 行)

轮盘上有 38 个空格。在这些空格中，18 个是黑色的，18 个是红色的，2 个是绿色的。绿色空格编号为 0 和 00。红色空格分别是 1、3、5、7、9、12、14、16、18、19、21、23、25、27、30、32、34 和 36。1 到 36 之间的其余整数用于对黑色空格编号。

许多不同的赌注可放在轮盘上。本练习只考虑以下部分：

- 单个数字(1~36、0 或 00)
- 红色和黑色
- 奇数和偶数(注意 0 和 00 不是偶数)

- 1 to 18 与 19 to 36

编写一个程序，使用 Python 的随机数生成器模拟一个旋转的轮盘赌轮。显示选中的号码和所有必须支付的赌注。例如，如果选择 13，那么程序应该显示：

```
The spin resulted in 13...
Pay 13
Pay Black
Pay Odd
Pay 1 to 18
```

如果仿真结果是 0 或 00，那么程序应该显示 Pay 0 或 Pay 00，没有任何进一步的输出。

第3章 循　　环

如何编写一个多次重复相同任务的程序？可复制代码并多次粘贴它，但是这样的解决方案并不优雅。它只允许任务执行固定次数，并需要对代码的每个副本进行增强或更正。

Python 提供了两个循环结构来克服这些限制。这两种类型的循环都允许程序中只出现一次的语句在程序运行时多次执行。如果有效地使用，循环可使用少量语句执行大量计算。

3.1　while 循环

while 循环导致一个或多个语句在条件的计算结果为 True 时执行。与 if 语句一样，while 循环有一个条件，该条件后面是一个缩进的主体。如果 while 循环的条件计算为 True，则执行循环体。当到达循环体的末尾时，执行流返回到循环的顶部，然后再次计算循环条件。如果条件仍然计算为 True，则循环体执行第二次。一旦执行流第二次到达循环体的末尾，执行流将再次返回到循环的顶部。循环的主体继续执行,直到 while 循环条件的计算结果为 False。发生这种情况时,将跳过循环的主体,并在 while 循环主体之后的第一个语句处继续执行。

许多 while 循环条件将变量保存的用户输入值与其他值进行比较。在循环体中读取值时,用户可通过输入适当的值来终止循环。具体地说,用户输入的值必须将 while 循环的条件赋值为 False。例如,下面的代码段读取用户输入的值,并报告每个值是正还是负。当用户输入 0 时,循环终止。在本例中没有显示任何消息。

```
# Read the first value from the user
x = int(input("Enter an integer (0 to quit): "))

# Keep looping while the user enters a non-zero number
while x != 0:
  # Report the nature of the number
  if x > 0:
    print("That's a positive number.")
  else:
    print("That's a negative number.")

  # Read the next value from the user
  x = int(input("Enter an integer (0 to quit): "))
```

这个程序首先从用户那里读取一个整数。如果该整数为 0,则 while 循环上的条件计算为 False。当这种情况发生时,跳过循环体,程序终止时不显示任何输出(除了输入提示)。如果 while 循环上的条件计算为 True,则执行循环体。

当循环体执行时,使用 if 语句将用户输入的值与 0 进行比较,并显示相应的消息。然后在循环的底部,读取用户输入的下一个值。因为已经到达循环的底部,所以控制流返回循环顶部,并再次计算它的条件。如果用户输入的最新值为 0,则条件的计算结果为 False。发生这种情况时,会跳过循环体,程序终止。否则,循环体将再次执行。它的主体继续执行,直到用户通过输入 0 将循环的条件赋值为 False 为止。

3.2 for 循环

与 while 循环一样，for 循环会导致只出现一次的语句在程序运行时执行多次。然而，用于确定这些语句执行多少次的机制对于 for 循环来说是非常不同的。

for 循环对集合中的每个项执行一次。集合可以是一组整数、字符串中的字母，或者存储在数据结构(如列表)中的值。for 循环的语法结构如下所示，其中<variable>、<collection>和<body>是必须适当填写的占位符。

```
for <variable> in <collection>:
  <body>
```

循环体由一个或多个可执行多次的 Python 语句组成。特别是，这些语句将对集合中的每个项执行一次。与 while 循环体一样，for 循环体也是缩进的。

在循环体执行集合中的每个项之前，该项将复制到<variable>中。这个变量是由 for 循环在执行时创建的。不需要使用赋值语句来创建它，在每次循环迭代开始时，可能覆盖之前分配给该变量的任何值。变量可以像其他 Python 变量一样在循环体中使用。

通过调用 Python 的 range()函数可以构造整数集合。使用一个参数调用 range()将返回一个范围，该范围从 0 开始，递增到该参数的值(不含)。例如，range(4)返回一个由 0、1、2 和 3 组成的范围。

给 range()提供两个参数时，返回的值集合从第一个参数增加到第二个参数(不含)。例如，range(4,7)返回一个由 4、5 和 6 组成的范围。当使用两个参数调用 range()并且第一个参数大于或等于第二个参数时，将返回一个空范围。每当将 for 循环应用于空范围时，都会跳过 for 循环体。执行流将继续执行 for 循环体之后的第一条语句。

还可使用第三个参数调用 range()函数，该参数是用于从范围内的初始值移动到最终值的步长值。使用大于 0 的步长值会产生一个范围，该范围从第一个参数开始，递增到第二个参数(不含)，每次递增步长值。使用负步长值允许构造递减值的集合。例如，当调用 range(0, -4)时，返回一个空范围；调用 range(0, -4, -1)返回一个由 0、-1、-2 和-3 组成的范围。请注意，作为第三个参数传递给 range 的步长值必须是一个整数。需要非整数步长值的问题通常用 while 循环来解决，而不是 for 循环。

下面的程序使用 for 循环和 range()函数来显示 3 的所有正倍数到用户输入的值(含)。

```
# Read the limit from the user
limit = int(input("Enter an integer: "))

# Display the positive multiples of 3 up to the limit
print("The multiples of 3 up to and including", limit, "are:")
for i in range(3, limit + 1, 3):
    print(i)
```

当这个程序执行时，首先从用户那里读取一个整数。假设用户输入 11，与后面描述这个程序其余部分的执行时输入的值相同。读取输入值后，执行流将继续执行描述程序输出的 print 语句。然后开始执行 for 循环。

整数的范围从 3 开始，一直到 11 + 1 = 12(不含)，每次递增 3。因此，范围包括 3、6 和 9。当循环第一次执行时，范围内的第一个整数赋值给 i，执行循环体，并显示 3。

一旦循环体第一次执行完毕，控制流返回循环的顶部和范围内的下一个值 6，它赋给 i。循环体再次执行并显示 6。然后，控制流再次返回循环的顶部。

下一个赋给 i 的值是 9。它在循环体下一次执行时显示。然后循环终止，因为范围内没有其他值。通常情况下，执行流在 for 循环体之后的第一个语句中继续。但是，这个程序中没有这样的语句，因此程序终止。

3.3 嵌套循环

循环体中的语句可以包含另一个循环。当发生这种情况时，内部循环被称为嵌套在外部循环中。任何类型的循环都可嵌套在任何其他类型的循环中。例如，下面的程序使用嵌套在 while 循环中的 for 循环，来重复用户输入的消息，直到用户输入空白消息为止。

```
# Read the first message from the user
message = input("Enter a message (blank to quit): ")

# Loop until the message is a blank line
while message != "":
  # Read the number of times the message should be displayed
  n = int(input("How many times should it be repeated? "))

  # Display the message the number of times requested
  for i in range(n):
    print(message)

  # Read the next message from the user
  message = input("Enter a message (blank to quit): ")
```

当执行这个程序时，首先读取来自用户的第一个消息。如果该消息不是空的，就执行 while 循环体，程序将读取用户输入消息的重复次数 n。整数的范围是从 0 到 n(不含)。然后 for 循环体打印消息 n 次，因为对于范围内的每个整数，消息都显示一次。

在 for 循环执行 n 次后，从用户那里读取下一条消息。然后执行流返回 while 循环的顶部，并对其条件进行计算。如果条件的计算结果为 True，则 while 循环体将再次运行。从用户处读取另一个整数，该整数覆盖之前的 n 值，然后 for 循环将消息打印 n 次。这将一直进行下去，直到 while 循环上的条件计算为 False 为止。当发生这种情况时，跳过 while 循环主体，程序终止，因为在 while 循环体之后没有语句可执行。

3.4 练习

下面的练习都应该用循环来完成。某些情况下，练习指定使用什么类型的循环。在其他情况下，必须自己做决定。有些练习可以用 for 循环和 while 循环轻松完成。其他练习更适合一种循环，而不适合另一种。此外，有些练习需要多个循环。当涉及多个循环时，一个循环可能需要嵌套在另一个循环中。在设计每个问题的解答时，要仔细考虑循环的选择。

练习 63 平均

(26 行)

本练习将创建一个程序，来计算用户输入的值集合的平均值。用户将输入 0 作为标记值，表示不再提供任何值。如果用户输入的第一个值是 0，则程序应该显示适当的错误消息。

> 提示：因为 0 表示输入的结束，所以它不应包含在平均值中。

练习 64：折扣表

(18 行)

某零售商正在以六折的价格出售多种已停产的产品。零售商希望帮助顾客确定商品的降价，方法是在货架上放一张打印好的折扣表，显示商品的原价和打折后的价格。编写一个程序，使用循环来生成这个表，显示原始价格、折扣金额和购买产品的新价格$4.95、$9.95、$14.95、$19.95 和$24.95。确保折扣金额和新价格在显示时四舍五入到小数点后两位。

练习 65：温度换算表

(22 行)

编写一个程序，显示摄氏温度和华氏温度转换表。该表的行应该包括 0 到 100 摄氏度之间的所有温度，这些温度是 10 摄氏度的倍数。列上适当的标题。摄氏度和华氏度之间的换算公式可在互联网上找到。

练习 66：不要再花钱了

(有解答，39 行)

2013 年 2 月 4 日是加拿大皇家铸币局发行硬币的最后一天。既然便士已经被淘汰，零售商必须调整总数，使它们在用现金支付时是 5 美分的倍数(信用卡和借记卡交易继续以 1 美分计)。虽然零售商在这方面有一定的自由度,但大多数人会选择四舍五入到最接近 5 美分的价格。

编写一个程序，从用户那里读取价格，直到输入空行为止。在一行显示所有输入项的总成本,在第二行显示客户用现金支付的应付金额。现金付款的金额应四舍五入到最接近的 5 美分。计算现金支付金额的一种方法是首先确定总共需要支付多少便士。然后计算总便士数除以 5 的余数。最后，如果余数小于 2.5，则向下调整总数。否则向上调整总数。

练习 67：计算多边形的周长

(有解答，42 行)

编写一个计算多边形周长的程序。首先从用户处读取多边形周长上的第一个点的 x 和 y 坐标。然后继续读取值对，直到用户为 x 坐标输入空行为止。每次读取一个额外坐标，都应该计算到前一个点的距离，并把它加到周长上。为 x 坐标输入空行时，程序应该将最后一点到第一点的距离加到周长上。然后显示周长。示例输入和输出值如下所示。用户输入的值以粗体显示。

```
Enter the first x-coordinate: 0
Enter the first y-coordinate: 0
Enter the next x-coordinate (blank to quit): 1
Enter the next y-coordinate: 0
```

```
Enter the next x-coordinate (blank to quit): 0
Enter the next y-coordinate: 1
Enter the next x-coordinate (blank to quit):
The perimeter of that polygon is 3.414213562373095
```

练习 68：计算平均绩点

(62 行)

练习 52 包括一个表，表中显示了某一特定学术机构从字母成绩到绩点的转换。本练习将计算用户输入的任意数量字母等级的平均绩点。用户输入一个空白行，表明所有分数都已提供。例如，如果用户输入 A，然后是 C+，然后是 B，然后是空行，那么程序应该报告的绩点平均值是 3.1。

完成这个练习时，会发现 52 题的答案很有帮助。程序不需要做任何错误检查。它可以假设用户输入的每个值都是有效的字母等级或空行。

练习 69：票价

(有解答，38 行)

某家动物园根据游客的年龄来决定门票价格。2 岁及以下的客人免费入场。3 至 12 岁的儿童的门票是 14.00 美元。65 岁以上的老年人的门票是 18.00 美元。其他所有游客的门票是 23 美元。

创建一个程序，首先从用户处读取组中所有游客的年龄，每行输入一个年龄。用户将输入一个空白行，以指示组中没有更多游客。然后，程序应该用适当消息显示该组的入场费。费用应该用两位小数来表示。

练习 70：奇偶校验位

(有解答，27 行)

奇偶校验位是一种简单机制，用于检测通过不可靠的连接(如电话线)传输的数据中的错误。基本思想是在每组 8 位之后传输一个额外的位，这样就可以检测传输中的单个位错误。

奇偶校验位既可计算为偶校验位,也可计算为奇校验位。如果选择偶校验位,则选择所传输的奇偶校验位,使所传输的"1"位(8位数据加上奇偶校验位)的总数为偶数。当奇数奇偶校验被选择时,奇偶校验位被选择,这样传送的"1"位的总数是奇数。

编写一个程序,使用偶数奇偶校验为用户输入的一组8位计算奇偶校验位。程序应该读取包含8位的字符串,直到用户输入空行为止。在用户输入每个字符串之后,程序应该显示一条明确的消息,指示奇偶校验位应该是0还是1。如果用户输入的不是8位,则显示适当的错误消息。

> **提示**:应该将来自用户的输入作为字符串读取。然后可以使用count方法来帮助确定字符串中0和1的数量。有关count方法的信息可在网上找到。

练习71:近似 π

(23 行)

π 的值可以用下面的无穷级数近似得到:

$$\pi \approx 3 + \frac{4}{2\times3\times4} - \frac{4}{4\times5\times6} + \frac{4}{6\times7\times8} - \frac{4}{8\times9\times10} + \frac{4}{10\times11\times12} - \cdots$$

编写一个程序,显示15个π的近似值。第一个近似应该只使用无穷级数的第一项。程序所显示的下一个近似值应包括无穷级数的更多一项,使它比以前的π近似值更接近π。

练习72:Fizz-Buzz

(17 行)

Fizz-Buzz 是一个游戏,有时是孩子们玩,以帮助他们学习除法。玩家通常围成一个圆圈,这样游戏就可以从一个玩家持续到另一个玩家。发球员先说"1",然后传给左边的球员。每个后续玩家在游戏传递给下一个玩家之前,依次说出下一个整数。轮到一名玩家时，他必须说出自己的数字或以下替换内容：
- 如果玩家的数字能被 3 整除，那么玩家说 fizz 而不是数字。
- 如果玩家的数字能被 5 整除，那么玩家说的是 buzz 而不是数字。

对于能被 3 和 5 整除的数，玩家必须同时说出 fizz 和 buzz。任何没有执行正确的替换或者在回答之前犹豫的玩家将被淘汰出局。最后一名玩家获胜。

编写一个程序，在 Fizz-Buzz 游戏中显示前 100 个数字的答案。每个答案应该单独占一行。

练习 73：凯撒密码

(有解答，35 行)

最早为人所知的加密例子之一是尤利乌斯·凯撒(Julius Caesar)使用的。凯撒需要向他的将军们提供书面指示，但他不想让敌人知道他的计划，为了避免信息落入敌人的手中，他发明了后来被称为凯撒密码的密码。

这个密码背后的思想很简单(同样，它也没有提供针对现代密码破解技术的保护)。原信息中的每个字母移动了 3 个位置。结果，A 变成了 D，B 变成了 E，C 变成了 F，D 变成了 G，等等。

字母表中的最后三个字母放在开头：X 变成 A，Y 变成 B，Z 变成 C。非字母字符不被密码修改。

编写一个实现凯撒密码的程序。允许用户提供消息和移位量，然后显示移位后的消息。确保程序同时编码大写和小写字母。程序还应该支持负移位值，以便它既可以用于编码消息，也可以用于解码消息。

练习 74：平方根

(14 行)

编写一个实现牛顿方法的程序，来计算和显示用户输入的数字

x 的平方根。牛顿法的算法如下：

> 读取用户输入的 x
> 初始化 guess 为 x /2
> 若 guess 还不够好
> > 将 guess 更新为 guess 和 x /guess 的平均值

当这个算法完成时，guess 包含了一个 x 平方根的近似值。近似值的质量取决于如何定义"足够好"。在本书的解答中，当 guess * guess 和 x 之差的绝对值小于或等于 10^{-12} 时，guess 就被认为是足够好了。

练习 75：字符串是回文吗？

(有解答，26 行)

如果一个字符串向前读取和向后读取的内容是相同的，那么它就是回文(palindrome)。例如 anna、civic、level 和 hannah 都是回文词。编写一个程序，从用户那里读取一个字符串，并使用循环来确定它是否是一个回文。显示结果，包括有意义的输出消息。

> 恐回文症(Aibohphobia)是对回文的极端或非理性的恐惧。这个词是由其反义词加上-phobia 后缀构成的，结果是一个回文。Ailihphilia 是对回文的喜爱。它是由-philia 后缀以相同的方式构成的。

练习 76：多个单词的回文

(35 行)

当忽略空格时，有许多短语是回文。例子包括"go dog""flee to me remote elf"和"some men interpret nine memos"等。扩展练习 75 的解答，以便在确定字符串是否是回文时忽略空格。对于另一个挑战，进一步扩展解答，这样也可以忽略标点符号，并将大写字母和小写字母视为等同。

练习 77：乘法表

(有解答，18 行)

本练习将创建一个显示乘法表的程序，该乘法表显示从 1 * 1 到 10 * 10 的所有整数组合的乘积。乘法表应该在表的顶部包括一排包含数字 1 到 10 的标签。它还应该在左边包括由数字 1 到 10 组成的标签。程序的预期输出如下所示：

```
      1   2   3   4   5   6   7   8   9  10
 1    1   2   3   4   5   6   7   8   9  10
 2    2   4   6   8  10  12  14  16  18  20
 3    3   6   9  12  15  18  21  24  27  30
 4    4   8  12  16  20  24  28  32  36  40
 5    5  10  15  20  25  30  35  40  45  50
 6    6  12  18  24  30  36  42  48  54  60
 7    7  14  21  28  35  42  49  56  63  70
 8    8  16  24  32  40  48  56  64  72  80
 9    9  18  27  36  45  54  63  72  81  90
10   10  20  30  40  50  60  70  80  90 100
```

当完成这个练习时，你可能发现，打印出一个值而不需要向下移动到下一行是很有帮助的。这可通过添加 end="" 作为 print 语句的最后一个参数来实现。例如，print("A") 将显示字母 A，然后向下移动到下一行。语句 print("A", end="") 将显示字母 A，而不向下移动到下一行，从而导致下一个 print 语句将其结果显示在与字母 A 相同的行上。

练习 78：科拉茨猜想

(22 行)

考虑这样一个整数序列：

从序列中唯一的正整数开始
while 若序列中的最后一项不等于 do
　if 最后一项是偶数 then
　　使用整数除法将最后一项除以 2，向序列中添加另一项

else

　　通过将最后一项乘以 3 再加 1，向序列中添加另一项

科拉茨猜想指出，当序列以任意正整数开始时，最终会以 1 结束。虽然这个推测从未被证实，但它似乎是正确的。

创建一个程序，从用户那里读取整数 n，并报告从 n 开始到 1 结束的序列中的所有值。程序应该允许用户继续输入新的 n 值(并且应该继续显示序列)，直到用户输入一个小于或等于零的 n 值。

科拉茨猜想是数学中一个开放问题的例子。虽然许多人试图证明这是真的，但没有人能做到。有关其他数学开放问题的信息可以在因特网上找到。

练习 79：最大公约数

(有解答，17 行)

两个正整数 n 和 m 的最大公约数是可以被 n 和 m 整除的最大数 d。有几种算法可用来解决这个问题，包括：

　　将 d 初始化为 m 和 n 中较小的那个。
　　while d 不能被 m 整除或者 d 不能被 n 整除　do
　　　　使 d 的值减 1
　　报告 d 是 n 和 m 的最大公约数

编写一个程序，从用户那里读取两个正整数，并使用这个算法来确定和报告它们的最大公约数。

练习 80：素因子

(27 行)

整数 n 的质因数分解可通过以下步骤确定：

　　初始化 factor 为 2
　　while factor 小于或等于 n　do

if n 能被 factor 整除 then
　　factor 是 n 的因子
　　用整数除法对 n 除以 factor
else
　　给 factor 加 1

编写一个程序，从用户那里读取一个整数。如果用户输入的值小于 2，则程序应该显示适当的错误消息。否则，程序应该显示可以相乘来计算 n 的质数，每行显示一个因子。例如：

```
Enter an integer (2 or greater): 72
The prime factors of 72 are:
2
2
2
3
3
```

练习 81：二进制转换到十进制

(18 行)

编写一个程序，将二进制数(以 2 为基数)转换为十进制数(以 10 为基数)。程序应该先以字符串形式读取用户输入的二进制数。然后通过处理二进制数中的每个数字，来计算等效的十进制数。最后，程序应该使用适当消息来显示等效的十进制数。

练习 82：十进制转换到二进制

(有解答，27 行)

编写一个程序，将一个十进制数(以 10 为基数)转换为二进制数(以 2 为基数)。将用户输入的十进制数作为整数读取，然后使用如下所示的除法算法执行转换。当算法完成时，结果包含数字的二进制表示。显示结果以及适当的消息。

设 result 为空字符串

设 q 为要转换的数字
repeat
　　设 r 等于 q 除以 2 的余数
　　将 r 转换为字符串，并将其添加到 result 的开头
　　将 q 除以 2，舍弃余数，并将结果存储回 q 中
until q = 0

练习 83：最大整数

(有解答，34 行)

本练习检查确定整数集合中最大值的过程。每个整数都从 1 到 100 之间的数字中随机选择。整数集合可能包含重复的值，也可能不包括一些介于 1 和 100 之间的整数。

花点时间想想如何在纸上解决这个问题。许多人会依次检查每个整数，然后问自己，他们目前考虑的数字是否比他们之前看到的最大的数字大。如果是，他们就会忘记之前的最大数字，而将当前数字记为新的最大数字。这是一种合理方法，如果仔细执行这个过程，就会得到正确答案。如果执行此任务，预计需要更新最大值和记住新数字的次数是多少？

虽然可以使用概率论来回答上一段末尾提出的问题，但我们通过模拟情境来探索它。创建一个程序，先选择 1 到 100 之间的随机整数，将该整数保存为到目前为止遇到的最大数字。选择初始整数后，生成 1 到 100 之间的另外 99 个随机整数。在生成每个整数时检查它，看看它是否大于目前遇到的最大数字。如果是，那么程序应该更新遇到的最大数字，并计算执行更新的次数。生成后显示每个整数。包含一个表示新的最大值的整数符号。

在显示 100 个整数之后，程序应该显示所遇到的最大值，以及在此过程中更新最大值的次数。程序的部分输出如下所示，…表示程序将显示的其余整数。运行程序几次。在最大值上执行的更新次数是否符合预期？

```
30
74 <== Update
58
17
40
37
13
34
46
52
80 <== Update
37
97 <== Update
45
55
73
...

The maximum value found was 100
The maximum value was updated 5 times
```

练习 84：抛硬币模拟

(47 行)

在连续三次抛硬币得到相同结果之前(三个都是正面或三个都是反面)，抛硬币的最少次数是多少？需要抛掷的最大次数是多少？平均需要多少次抛掷？这个练习将创建一个程序,模拟一系列抛硬币游戏来探索这些问题。

创建一个程序，使用 Python 的随机数生成器模拟抛硬币几次。模拟的硬币应该是公平的，这意味着正面的概率等于反面的概率。程序应该投掷模拟的硬币，直到出现 3 个连续的正面和 3 个连续的反面。每次结果为正面时显示一个 H，每次结果为反面时显示一个 T，模拟的所有结果都在同一行上。然后显示连续出现 3 次相同结果所需的投掷次数。当程序运行时，它应该执行模拟 10 次，并报告所需的平均投掷次数。样本输出如下所示：

```
H T T T (4 flips)
H H T T H T H T T H H T H T T H T T T (19 flips)
T T T (3 flips)
```

第3章 循　环

```
T H H H (4 flips)
H H H (3 flips)
T H T T H T H H T T H H T H T H H H (18 flips)
H T T H H H (6 flips)
T H T T T (5 flips)
T T H T T H T H T H H H (12 flips)
T H T T T (5 flips)
On average, 7.9 flips were needed.
```

第4章 函 数

随着我们编写的程序长度的增加,需要设法使它们更容易开发和调试。一种方法是将程序代码分解成函数。

函数有几个重要用途:允许编写一次代码,然后从多个位置调用它;允许分别测试解答的不同部分;允许在完成程序的一部分后隐藏细节。为实现这些目标,函数允许程序员命名并留出一组Python语句,供以后使用。然后,程序可在需要时执行这些语句。这些语句通过定义一个函数来命名。这些语句是通过调用一个函数来执行的。当函数中的语句执行完成时,控制流返回到调用函数的位置,程序继续从该位置执行。

以前编写的程序调用 print()、input()、int()和 float()等函数。所有这些函数都是由创建 Python 编程语言的人定义的,这些函数可以在任何 Python 程序中调用。在本章中,除了调用前面定义的函数外,还将学习如何定义和调用自己的函数。

函数定义以一行 def 开始,后跟要定义的函数的名称,后面是一个左括号、一个右括号和一个冒号。这一行后面是函数的主体,它是调用函数时将执行的语句集合。与 if 语句和循环的主体一样,函数的主体也是缩进的。若下一行的缩进量与以 def 开头的行相同或更小，则函数的主体在该行之前结束。例如，下面的代码行定义了一个函数，它用星号字符绘制了一个方框。

```
def drawBox():
  print("***********")
  print("*         *")
  print("*         *")
  print("***********")
```

单独而言，这些代码行不会产生任何输出，因为虽然定义了 drawBox()函数，但它从未被调用。定义函数会将这些语句放在一边，以备将来使用，并将名称 drawBox()与它们关联起来，但不会执行它们。只包含这些行的 Python 程序是有效的程序，但在执行时不会生成任何输出。

调用 drawBox()函数的方法是使用它的名称，后跟一个开括号和一个闭括号。将下面的代码行添加到前一个程序的末尾(不缩进它)，就调用该函数并绘制方框。

```
drawBox()
```

添加该行的第二个副本将导致绘制第二个方框，添加该行的第三个副本将导致绘制第三个方框。更一般地说，在解题时，可以根据需要多次调用一个函数，这些调用可以在程序中的许多不同位置进行。每次调用函数时，函数体中的语句都会执行。当函数返回时，就执行紧跟在函数调用后面的语句。

4.1 带参数的函数

drawBox()函数工作正常。它绘制了预期要绘制的特定方框，但是它并不灵活，因此并没有发挥应有的作用。特别是，如果该函数可以绘制许多不同大小的方框，它将更加灵活和有用。

第4章 函　　数

　　许多函数的参数是在调用函数时在括号内提供的值。当定义函数时，函数在参数变量中接收这些参数值，这些参数变量包含在括号中。函数定义中的参数变量数量指示调用函数时必须提供的参数数量。

　　通过向 drawBox()函数的定义添加两个参数，可以使它变得更有用。这些由逗号分隔的参数分别保存方框的宽度和高度。函数体使用参数变量中的值来绘制方框，如下所示。如果提供给函数的参数无效，则使用 if 语句和 quit()函数立即结束程序。

```
## Draw a box outlined with asterisks and filled with spaces.
# @param width the width of the box
# @param height the height of the box
def drawBox(width, height):
  # A box that is smaller than 2x2 cannot be drawn by this function
  if width < 2 or height < 2:
    print("Error: The width or height is too small.")
    quit()

  # Draw the top of the box
  print("*" * width)

  # Draw the sides of the box
  for i in range(height - 2):
    print("*" + " " * (width - 2) + "*")

  # Draw the bottom of the box
  print("*" * width)
```

　　调用 drawBox()函数时必须提供两个参数，因为它的定义包含两个参数变量。当函数执行时，第一个参数的值放在第一个参数变量中，类似地，第二个参数的值放在第二个参数变量中。例如，下面的函数调用绘制一个宽度为 15 个字符、高度为 4 个字符的方框。通过使用不同的参数再次调用该函数，可以绘制不同大小的其他方框。

```
drawBox(15,4)
```

　　在当前形式中，drawBox()函数总是用星号字符绘制方框的轮廓，并总是用空格填充方框。虽然这在很多情况下都可很好地工作，但有时程序员也需要用不同字符绘制或填充的方框。为此，更新

drawBox()，使它接收两个额外的参数，分别指定轮廓和填充字符。还必须更新函数主体，以使用这些额外的参数变量，如下所示。对 drawBox() 函数的调用包含在程序末尾处，该函数用@符号勾勒出方框，并用句点填充方框。

```python
## Draw a box.
# @param width the width of the box
# @param height the height of the box
# @param outline the character used for the outline of the box
# @param fill the character used to fill the box
def drawBox(width, height, outline, fill):
    # A box that is smaller than 2x2 cannot be drawn by this function
    if width < 2 or height < 2:
        print("Error: The width or height is too small.")
        quit()

    # Draw the top of the box
    print(outline * width)

    # Draw the sides of the box
    for i in range(height - 2):
        print(outline + fill * (width - 2) + outline)

    # Draw the bottom of the box
    print(outline * width)

# Demonstrate the drawBox function
drawBox(14, 5, "@", ".")
```

每次调用这个版本的 drawBox() 时，程序员必须包括轮廓字符和填充字符(除了宽度和高度之外)。虽然在某些情况下可能需要这样做，但是当星号和空格比其他字符组合使用得更频繁时，就会令人沮丧，因为每次调用函数时都必须重复这些参数。为解决这个问题，将轮廓字符和填充字符的默认值添加到函数的定义中。参数的默认值与名称之间用等号分隔，如下所示。

```python
def drawBox(width, height, outline="*", fill=" "):
```

一旦进行了这种更改，可以用两个、三个或四个参数调用 drawBox()。如果使用两个参数调用 drawBox()，则第一个参数放在 width 参数变量中，第二个参数放在 height 参数变量中。outline 和

fill 参数变量将分别保存其默认值：星号和空格。使用这些默认值是因为在调用函数时，没有为这些参数提供值。

现在考虑以下对 drawBox() 的调用：

```
drawBox(14, 5, "@", ".")
```

这个函数调用包含四个参数。前两个参数是 width 和 height，它们放在这些参数变量中。第三个参数是 outline 字符。因为已经提供了它，所以默认的 outline 值(星号)被替换为提供的值，它是一个 @ 符号。类似地，因为调用包含第四个参数，所以默认的 fill 值将被替换为句点。前面对 drawBox() 的调用产生的结果如下所示。

```
@@@@@@@@@@@@@@
@............@
@............@
@............@
@@@@@@@@@@@@@@
```

4.2 函数中的变量

在函数内部创建变量时，该变量是该函数的局部变量。这意味着变量只在函数执行时存在，且只能在函数体中访问它。当函数返回时，变量将不复存在，因此，在此之后将无法访问该变量。drawBox() 函数使用多个变量来执行其任务。这些变量包括在调用函数时创建的参数变量(如 width 和 fill)，以及在开始执行循环时创建的 for 循环控制变量 i。所有这些都是只能在这个函数中访问的局部变量。在函数体中使用赋值语句创建的变量也是局部变量。

4.3 返回值

方框绘制函数在屏幕上打印字符。虽然它使用形参来指定如何绘制方框,但该函数不计算需要存储在变量中并在程序后面使用的结果。但很多函数都计算这样的值。例如,math 模块中的 sqrt 函数计算其参数的平方根并返回此值,以便在后续计算中使用它。类似地,input()函数读取用户输入的值,然后返回,以便以后在程序中使用。用户编写的一些函数也需要返回值。

函数在返回值时,使用 return 关键字,后跟要返回的值。当 return 执行时,函数立即结束,控制流返回到调用函数的位置。例如,下面的语句立即结束函数的执行,并将 5 返回到调用它的位置。

```
return 5
```

有返回值的函数通常在赋值语句的右侧调用,也可以在需要值的其他上下文中调用。这种情形包括 if 语句或 while 循环条件,或作为其他函数(如 print()或 range())的参数。

不返回结果的函数不需要使用 return 关键字,因为函数会在函数体的最后一个语句执行之后自动返回。但是,程序员可以使用 return 关键字(不带尾随值)来强制函数返回其主体中较早的位置。任何函数,不管是否返回值,都可以包含多个 return 语句。这样的函数在任何一个 return 语句执行后立即返回。

考虑下面的例子。等比数列是一组项的序列,这些项以某个值 a 开始,然后是无穷多个附加项。除了第一项之外,数列中的每一项都是由它的直接前项乘以 r 计算出来的,r 被称为公比。结果,序列中的项是 a、ar、ar^2、$ar^3\cdots$,当 r 为 1 时,等比数列的前 n 项之和为 $a \times n$,当 r 不为 1 时,等比数列的前 n 项之和可由下式计算:

$$\text{sum} = \frac{a(1-r^n)}{1-r}$$

可以编写一个函数来计算任何等比序列的前 n 项之和。它需要 3 个参数:a、r 和 n,它需要返回一个结果,即前 n 项的和。该函数的代码如下所示。

第4章 函　　数

```
## Compute the sum of the first n terms of a geometric sequence.
# @param a the first term in the sequence
# @param r the common ratio for the sequence
# @param n the number of terms to include in the sum
# @return the sum of the first n term of the sequence
def sumGeometric(a, r, n):

    # Compute and return the sum when the common ratio is 1
    if r == 1:
        return a * n

    # Compute and return the sum when the common ratio is not 1
    s = a * (1 - r ** n) / (1 - r)

    return s
```

函数首先使用 if 语句来判断 r 是否为 1。如果是，则计算和为 *n，函数立即返回该值，而不执行函数主体中的其余代码行。当 r 不等于 1 时，跳过 if 语句的主体，计算前 n 项的和并将其存储在 s 中。然后将存储在 s 中的值返回到调用函数的位置。

下面的程序演示了 sumGeometry() 函数，它会计算总和，直到用户为 a 输入 0。每个总和在函数中计算，然后返回到调用函数的位置。接着使用赋值语句将返回值存储在 total 变量中。后续的语句显示 total，之后程序继续运行，从用户处读取另一个序列的值。

```
def main():
    # Read the initial value for the first sequence
    init = float(input("Enter the value of a (0 to quit): "))

    # While the initial value is non-zero
    while init != 0:
        # Read the ratio and number of terms
        ratio = float(input("Enter the ratio, r: "))
        num = int(input("Enter the number of terms, n: "))

        # Compute and display the total
        total = sumGeometric(init, ratio, num)
        print("The sum of the first", num, "terms is", total)

        # Read the initial value for the next sequence
        init = float(input("Enter the value of a (0 to quit): "))

# Call the main function
main()
```

4.4 将函数导入其他程序

使用函数的好处之一是函数只需要编写一次，就可以从不同位置多次调用它。当函数定义和调用位置都位于同一文件中时，这很容易实现。先定义函数，调用时只需要使用它的名称，后跟包含参数的括号。

在某些情况下解决新问题时，希望调用为前一个程序编写的函数。新程序员(甚至一些有经验的程序员)常常试图将函数从包含旧程序的文件复制到包含新程序的文件中，但这是一种不可取的方法。复制函数的结果是相同的代码驻留在两个地方。因此，当一个错误被识别时，它需要被修正两次。更好的方法是把函数从旧程序导入新程序中，类似于从 Python 内置模块中导入函数的方式。

为将旧 Python 程序中的函数导入新程序中，可以使用 import 关键字，然后是包含相关函数的 Python 文件的名称(没有.py 扩展名)。这允许新程序调用旧文件中的所有函数，但也会导致执行旧文件中的程序。虽然在某些情况下这可能是可取的，但通常需要在不实际运行程序的情况下访问旧程序的函数。通常的实现方式是：创建一个名为 main()的函数，该函数包含解决问题所需的语句。然后，文件末尾的一行代码调用 main()函数。最后，添加一个 if 语句，以确保当文件导入到另一个程序中时，main()函数不会执行，如下所示：

```
if __name__ == "__main__":
    main()
```

当创建的程序包含将来可能要导入到另一个程序中的函数时，应该使用这种结构。

4.5 练习

函数允许命名 Python 语句序列，并从程序中的多个位置调用它们。与不定义任何函数的程序相比，这提供了一些优势，包括代码可以编写一次。却在多个位置调用，以及可以单独测试解答的不同部分。函数还允许程序员在专注于解答的其他方面时，将程序的一些细节放在一边。有效地使用函数将有助于编写更好的程序，特别是在处理更大的问题时。函数应该在完成本章的所有练习时使用。

练习 85：计算斜边

(23 行)

编写一个函数，以直角三角形的两个直角边的长度为参数。返回使用勾股定理计算的三角形的斜边作为函数的结果。包含一个 main 程序，它从用户那里读取直角三角形中直角边的长度，使用函数计算斜边的长度，并显示结果。

练习 86：出租车费

(22 行)

在一个特定的司法管辖区，出租车费包括基本车费$4.00，每行驶 140 米另加$0.25。编写一个函数，该函数将旅行距离(以公里为单位)作为唯一参数，并返回总车费作为唯一结果。编写一个 main 程序来演示这个函数。

> 提示：出租车票价会随时间变化。使用常量来表示基本票价和票价的可变部分，以便在费率增加时可以方便地更新程序。

练习 87：运费计算器

(23 行)

在线零售商为其许多商品提供快递服务,订单中第一项商品的运费为 10.95 美元，同一订单中后续每项商品的运费为 2.95 美元。编写一个函数，以商品的数量作为其唯一的参数。将订单的运费作为函数的结果返回。包括一个 main 程序，读取从用户购买的物品数量，并显示运费。

练习 88：三个值的中位数

(有解答，43 行)

编写一个函数，以三个数字为参数，并返回这些参数的中值作为其结果。包含一个 main 程序，从用户读取三个值，并显示它们的中位数。

> 提示：中值是按升序排列的三个值的中间值。可以使用 if 语句，或者使用一点数学创造力来找到它。

练习 89：将整数转换为序数

(47 行)

像第一、第二和第三这样的词被称为序数。本练习将编写一个函数，该函数的唯一参数为整数，并返回一个包含对应英文序号的字符串作为唯一结果。函数必须处理 1 到 12(含)之间的整数。如果函数调用的参数不在这个范围内，那么它应该返回一个空字符串。包含一个 main 程序，显示从 1 到 12 的每个整数及其序号来演示函数。main 程序应该只在文件没有被导入其他程序时运行。

练习 90：圣诞节的 12 天

(有解答，52 行)

《圣诞节的 12 天》是一首重复诵唱的歌曲，它描述了在每 12 天里送给真爱的礼物清单越来越长。第一天只送一份礼物。每增加一天，就把一个新礼物添加到清单中，然后完整的清单就会被发送出去。这首歌的前三段如下。完整歌词可在互联网上找到。

On the first day of Christmas

my true love sent to me:
A partridge in a pear tree.

On the second day of Christmas
my true love sent to me:
Two turtle doves,
And a partridge in a pear tree.

On the third day of Christmas
my true love sent to me:
Three French hens,
Two turtle doves,
And a partridge in a pear tree.

编写一个程序,显示《圣诞节的 12 天》的完整歌词。程序应该包含一个函数来显示歌词。它将把段号作为唯一的参数。然后程序应该用从 1 到 12 的整数调用这个函数 12 次。

歌曲中发送给收件人的每个条目应该只在程序中出现一次,鹧鸪可能除外。如果这能帮助处理第一段"A partridge in a pear tree"和第二段"And a partridge in a pear tree"之间的区别,它可能会出现两次。将解答导入练习 89,以帮助完成此练习。

练习 91:从公历日期到序数日期

(72 行)

序数日期包括年份和日期,这两个都是整数。年份可以是公历中的任意一年,一年中的日期从 1(表示 1 月 1 日)到 365(如果这一年是闰年,则是 366),表示 12 月 31 日。在计算日期之间相差的天数(而不是月份)时,序数日期非常方便。例如,序数日期可以很容易地确定客户是否在 90 天的退货期内,食品根据其生产日期确定是否在保质期内。

编写一个名为 ordinalDate() 的函数,该函数接收三个整数作为参数。这些参数分别为日、月和年。函数应该返回该日期在一年中的序数日期作为唯一结果。创建一个 main 程序,它读取用户的日期、月份和年份,并显示该日期在一年中的序数日期。main 程序应该只在文件没有被导入其他程序时运行。

练习 92：从序数日期到公历日期

(103 行)

创建一个函数，该函数使用一个序数日期作为参数，该日期由该年中的年份和日期组成。该函数返回与该序数日期对应的公历日期和月份作为结果。确保函数正确地处理闰年。

使用该函数以及之前编写的 ordinalDate() 函数，创建一个程序，读取用户输入的日期。然后，程序应该报告几天后的第二个日期。例如，程序可以阅读产品购买日期，然后报告可以退货的最后日期(基于退货期限，这是一个特定的天数)，或者程序可以根据妊娠期的 280 天计算婴儿的预产期。确保程序正确处理输入日期和计算日期出现在不同年份的情况。

练习 93：在终端窗口中居中一个显示字符串

(有解答，29 行)

编写一个函数，以字符串 s 作为第一个参数，以窗口的字符宽度 w 作为第二个参数。函数将返回一个新字符串，其中包含所需的前导空格，以便在打印时新字符串 s 在窗口中居中显示。新字符串的构造方法如下：

- 如果 s 的长度大于或等于窗口的宽度，那么应该返回 s。
- 如果 s 的长度小于窗口的宽度，则应该返回一个字符串，其中包含 (len(s) - w) // 2 个空格，后跟 s。

编写一个 main 程序，通过在窗口中显示多个居中的字符串来演示函数。

练习 94：它是一个有效的三角形吗？

(33 行)

如果有 3 根吸管，它们的长度可能不同，可以把它们放下来，当它们的端部接触时，就可能会形成一个三角形，也可能不会。例如，如果所有的吸管都有 6 英寸长，就很容易用它们构建一个等边三角形。但是，如果一根吸管有 6 英寸长，而另外两根只有 2 英寸长，就不能形成三角形。更一般地说，如果任意一根吸管的长度大于或等于另外两根吸管的长度之和，就构不成三角形。否则它们会形成一个三角形。

编写一个函数，来确定三个长度是否可以构成一个三角形。该函数接收 3 个参数，返回一个布尔结果。如果任何长度小于或等于 0，那么函数应该返回 False。否则，它应该确定长度是否可以使用前一段描述的方法来形成一个三角形，并返回适当结果。此外，编写一个程序，从用户那里读取 3 个长度，并演示函数的行为。

练习 95：大写

(有解答，68 行)

许多人不能正确地使用大写字母，尤其是在像智能手机这样的小设备上打字。为了帮助解决这种情况，请创建一个函数，该函数将字符串作为唯一的参数，并返回已正确大写的字符串的新副本。具体而言，函数必须：

- 大写字符串中的第一个非空格字符，
- 在句号、感叹号或问号后面的第一个非空格字符要大写
- 如果 i 前面有空格，后面有空格、句号、感叹号、问号或撇号，小写的 i 要大写。

实现这些转换将纠正大多数大小写错误。例如，如果给函数提供了字符串 "what time do i have to be there? what's the address? this time I'll try to be on time!"，就应该返回字符串 "What time do I have to be there? What's the address? This time I'll try to be on time!" 包含一个 main 程序，它从用户那里读取一个字符串，使用函数将其改为大写，并显示结果。

练习 96：字符串是否表示整数？

(有解答，30 行)

本练习将编写一个名为 isInteger 的函数，用于确定字符串中

的字符是否代表有效整数。确定字符串是否表示整数时，则应忽略开头或结尾的任何空白。一旦这个空白被忽略，如果字符串的长度至少是1，且只包含数字，则该字符串就表示一个整数，或者如果它的第一个字符是+或-，且第一个字符后面有一个或多个字符，则所有字符都是数字。

编写一个main程序，从用户那里读取一个字符串，并报告它是否代表一个整数。如果包含解答的文件被导入另一个程序中，请确保main程序不会运行。

> **提示**：在完成这个练习时，字符串的 lstrip、rstrip 和/或 strip 方法非常有用。这些方法的文档可在网上找到。

练习 97：操作符的优先级

(30 行)

编写一个名为 priority 的函数，该函数返回一个整数，该整数表示数学运算符的优先级。包含运算符的字符串作为函数的唯一参数传递给函数。函数应该给+和-返回1，给*和/返回2，给^返回3。如果传递给函数的字符串不是这些操作符中的一个，那么函数应该返回-1。包含一个 main 程序，该 main 程序从用户那里读取操作符，并显示操作符的优先级，或者显示一条表明输入不是操作符的错误消息。main 程序应该只在包含解答的文件未被导入其他程序时运行。

> 在这个练习以及本书后面的其他练习中，都使用^代表取幂。用^替代 Python 选项**将使这些练习更容易，因为操作符永远是一个字符。

练习 98：一个数是素数吗？

(有解答，28 行)

质数是大于 1 的整数，它只能被 1 和它本身整除。编写一个函数，来确定它的参数是不是质数，如果是，则返回 True，否则返回 False。编写一个 main 程序，从用户那里读取一个整数并显示一条消息，指示它是不是质数。如果包含解答的文件被导入另一个程序中，请确保 main 程序不会运行。

练习 99：下一个素数

(27 行)

这个练习将创建一个名为 nextPrime()的函数，该函数查找并返回第一个大于某个整数 n 的素数。n 的值作为函数的唯一参数传递给该函数。包含一个 main 程序，它从用户那里读取一个整数，并显示第一个大于输入值的素数。在完成本练习时，导入并使用该解答来完成练习 98。

练习 100：随机密码

(有解答，33 行)

编写一个生成随机密码的函数。密码的长度应该在 7 到 10 个字符之间。每个字符应该从 ASCII 表的第 33 位到第 126 位中随机选择。函数不接收任何参数，返回随机生成的密码作为唯一结果。在文件的 main 程序中显示随机生成的密码。main 程序应该只在解答没有被导入另一个文件时运行。

> 提示：在完成这个练习时，chr()函数很有帮助。关于这个函数的详细信息可在网上找到。

练习 101：随机车牌

(45 行)

在一个特定的司法管辖区,旧的车牌由三个字母和三个数字组成。当所有遵循该模式的牌照被使用后，车牌的格式更改为四个数字后面跟着三个字母。

编写一个函数，生成随机的车牌。函数为旧车牌或新车牌生成字符序列的概率应该大致相等。编写一个 main 程序，调用函数并显示随机生成的车牌。

练习 102：检查密码

(有解答，40 行)

本练习中将编写一个函数来确定密码是否正确。需要定义一个好的密码，它至少 8 个字符长，包含至少一个大写字母、至少一个小写字母和至少一个数字。如果作为唯一参数传递的密码正确，则函数应该返回 True。否则应该返回 False。包括一个 main 程序，从用户那里读取密码，并报告其是否正确。确保 main 程序只在解答没有导入另一个文件时运行。

练习 103：随机的好密码

(22 行)

使用练习 100 和 102 的解答，编写一个程序，生成一个随机的有效密码并显示它。计数和显示在生成一个有效密码之前需要尝试的次数。构造解答，以便它导入以前编写的函数，然后在为本练习创建的文件的 main() 函数中调用函数。

练习 104：十六进制和十进制数字

(41 行)

编写两个函数 hex2int() 和 int2hex()，在十六进制数字(0、1、2、3、4、5、6、7、8、9、A、B、C、D、E 和 F)和十进制(以 10 为基数)之间进行转换。hex2int() 函数负责将包含一个十六进制数字的字符串转换为基数为 10 的整数，而 int2hex() 函数负责将 0 到 15 之间的一个整数转换为十六进制数字。每个函数都把要转换的值作为唯一参数，并返回转换后的值作为唯一结果。确保 hex2int() 函数对于大小写字母都能正确工作。函数应该显示有意义的错误消息，如果参数值超出预期范围，则结束程序。

第 4 章 函　　数

练习 105：任意进制之间的转换

(有解答，71 行)

编写一个程序，允许用户将一个进制的数字转换成另一个进制。程序应该为输入数字和结果数字支持 2 到 16 之间的进制。如果用户选择此范围之外的进制，则应显示适当的错误消息，程序应退出。将程序分成几个函数，包括从任意进制转换为 10 进制的函数，从 10 进制转换为任意进制的函数，从用户那里读取进制和输入数字的 main 程序。完成这个练习时，会发现练习 81、练习 82 和练习 104 的答案是有帮助的。

练习 106：一个月的天数

(47 行)

编写一个函数来确定一个月有多少天。函数接收两个参数：月份，是 1 到 12 之间的整数；年份，是四位整数。确保函数报告了闰年 2 月的正确天数。包含一个 main 程序，它读取用户输入的月份和年份，并显示该月的天数。解答这个练习时，练习 58 的方法是有帮助的。

练习 107：最简分数

(有解答，46 行)

编写一个函数，它以两个正整数为参数，这两个正整数代表一个分数的分子和分母。函数体应将分数改为最简式，然后返回最简分数的分子和分母作为结果。例如，如果传递给函数的参数是 6 和 63，那么函数应该返回 2 和 21。包括一个 main 程序，允许用户输入分子和分母。然后程序应该显示简化的分数。

> **提示**：练习 79 编写了一个程序，来计算两个正整数的最大公约数。在完成这个练习时，练习 79 的代码是有用的。

练习 108：减少度量单位

(有解答，87 行)

许多食谱书仍然使用杯、汤匙和茶匙来度量烹饪或烘焙时使用的配料量。如果有合适的量杯和勺子，这些食谱很容易遵循，但在为整个大家庭做圣诞晚餐时，它们可能很难翻倍，或增加到三倍或四倍。例如，一个配方需要 4 汤匙的一种配料，当它翻了两番时，需要 16 汤匙。然而，16 汤匙的量以 1 杯形式表示会更好(也更容易测量)。

编写一个函数，用最大可能的单位表示一个特等品的体积。函数的第一个参数是单位的数量，第二个参数是度量单位(杯、汤匙或茶匙)。它将返回一个字符串，该字符串表示使用最大可能的单位来度量，作为其唯一结果。例如，如果函数的参数代表 59 茶匙，那么它应该返回字符串 "1 杯，3 汤匙，2 茶匙"。

提示：一杯相当于 16 汤匙。一汤匙相当于三茶匙。

练习 109：神奇的日子

(有解答，26 行)

对于神奇的日子，其日子乘以月份等于两位数的年份。例如，1960 年 6 月 10 日就是一个神奇的日子，因为 6 乘以 10 等于 60，即两位数字的年份。编写一个函数，来确定某日子是不是神奇的日子。使用函数创建 main 程序，该 main 程序查找并显示 20 世纪的所有神奇日子。完成这个练习时，可能会发现练习 106 的答案很有帮助。

第5章 列　　表

前面创建的每个变量都有一个值。值可以是整数、布尔值、字符串或其他类型的值。为每个值使用一个变量对于小问题来说是可行的，但在处理大量数据时，它很快就会变得站不住脚。列表允许在一个变量中存储多个值，帮助解决了这个问题。

保存列表的变量是使用赋值语句创建的,这与之前创建的变量很相似。列表用方括号括起来，逗号用于分隔列表中的相邻值。例如，下面的赋值语句创建一个包含 4 个浮点数的列表，并将其存储在一个名为 data 的变量中。然后通过调用 print()函数来显示这些值。当 print()函数执行时，所有 4 个值都会显示出来，因为 data 是值的整个列表。

```
data = [2.71, 3.14, 1.41, 1.62]
print(data)
```

列表可包含零个或多个值。空列表中没有值，用[]表示(一个左方括号紧接一个右方括号)。整数变量可以初始化为 0，然后在程序的后面给它增加值，同样，列表可以初始化为空列表，然后在程序执行时将项添加到其中。

5.1 访问单个元素

列表中的每个值都称为元素。列表中的元素以整数顺序编号，从 0 开始。每个整数标识列表中的特定元素，称为该元素的索引。在前面的代码段中，data 中索引 0 处的元素是 2.71，而索引 3 处的元素是 1.62。

为访问单个列表元素，可使用列表的名称，紧接用方括号括起来的元素索引。例如，下面的语句使用这个符号来显示 3.14。注意，在打印索引 1 处的元素，将显示列表中的第二个元素，因为列表中第一个元素的索引为 0。

```
data = [2.71, 3.14, 1.41, 1.62]
print(data[1])
```

可使用赋值语句更新单个列表元素。把列表的名称后跟用方括号括起来的元素索引，放在赋值操作符的左侧。将存储在该索引处的新值放在赋值操作符的右侧。当赋值语句执行时，先前存储在指定索引处的元素将被新值覆盖。列表中的其他元素不受此更改的影响。

考虑下面的例子。它创建一个包含四个元素的列表，然后用 2.30 替换索引 2 处的元素。当 print 语句执行时，它将显示列表中的所有值。这些值分别是 2.71、3.14、2.30 和 1.62。

```
data = [2.71, 3.14, 1.41, 1.62]
data[2] = 2.30
print(data)
```

5.2 循环和列表

for 循环对集合中的每个项执行一次。集合可以是通过调用 range() 函数构造的整数范围。它也可以是一个列表。下面的示例使用 for

循环来合计 data 中的值。

```
# Initialize data and total
data = [2.71, 3.14, 1.41, 1.62]
total = 0

# Total the values in data
for value in data:
  total = total + value

# Display the total
print("The total is", total)
```

这个程序首先将 data 和 total 初始化为所示的值。然后，for 循环开始执行。data 中的第一个值复制到 value 中，然后运行循环体。它把 value 加到 total 上。

一旦第一次执行循环体，控制流返回到循环的顶部。data 中的第二个元素复制到 value 中，循环体再次执行，将这个新值添加到总数中。循环对列表中的每个元素执行一次，并计算所有元素的总数，之后结束这个过程。然后显示结果，程序终止。

有时会构造循环来迭代列表的索引，而不是它的值。要构造这样的循环，需要能够确定列表中有多少元素。这可使用 len() 函数来完成。它需要一个参数，即列表，并返回列表中的元素数量[1]。

len() 函数可与 range() 函数一起用于构造一个整数集合，其中包含列表的所有索引。这是通过将列表的长度作为 range() 的唯一参数来实现的。索引的一个子集可以通过向 range() 提供第二个参数来构造。下面的程序演示了这一点，它使用 for 循环遍历 data 的所有索引(第一个索引除外)，以确定 data 中最大元素的位置。

```
# Initialize data and largest pos
data = [1.62, 1.41, 3.14, 2.71]
largest_pos = 0

# Find the position of the largest element
for i in range(1, len(data)):
  if data[i] > data[largest_pos]:
```

---

1 如果传递给 len() 函数的列表为空，则 len() 函数返回 0。

```
        largest_pos = i

# Display the result
print("The largest value is", data[largest_pos], \
      "which is at index", largest_pos)
```

这个程序首先初始化 data 和 largest_pos 变量。然后使用 range() 函数构造 for 循环将使用的值集合。它的第一个参数是 1，第二个参数是 data 的长度 4。因此，range 返回从 1 到 3(含)的连续整数的集合，这也是 data 中除第一个元素外的所有元素的索引。

for 循环通过将 1 存储到 i 中开始执行，然后循环体第一次运行。它将索引 i 处的 data 值与索引 largest_pos 处的数据值进行比较。由于索引 i 处的元素更小，if 语句的条件计算为 False，并跳过 if 语句的主体。

现在，控件流返回到循环的顶部。范围内的下一个值 2 存储到 i 中，循环体第二次执行。将索引 i 处的值与索引 largest_pos 处的值进行比较。由于索引 i 的值更大，所以执行 if 语句的主体，并且 largest_pos 设置为 i，即 2。

当 i 等于 3 时，循环再运行一次。索引 i 处的元素小于索引 largest_pos 处的元素，因此跳过 if 语句的主体。然后循环终止，程序报告最大值为 3.14，即索引 2。

while 循环也可用于处理列表。例如，下面的代码段使用 while 循环来标识列表中第一个正数的索引。循环使用一个变量 i，它保存列表中元素的索引，索引从 0 开始。i 中的值随着程序的运行而增加，直至到达列表的末尾或找到一个正数元素。

```
# Initialize data
data = [0, -1, 4, 1, 0]

# Loop while i is a valid index and the value at index i is not a positive value
i = 0
while i < len(data) and data[i] <= 0:
  i = i + 1

# If i is less than the length of data then the loop terminated because a positive number was
# found. Otherwise i will be equal to the length of data, indicating that a positive number
```

```
# was not found.
if i < len(data):
  print("The first positive number is at index", i)
else:
  print("The list does not contain a positive number")
```

当这个程序执行时，它首先初始化 data 和 i，然后计算 while 循环的条件。i 的值是 0，小于 data 的长度，而位置 i 的元素是 0，小于等于 0。结果，条件的值为 True，循环体执行，i 的值从 0 增加到 1。

控制流返回到 while 循环的顶部，再次计算其条件。存储在 i 中的值仍然小于 data 的长度，列表中位置 i 的值仍然小于或等于 0。结果，循环条件仍然计算为 True。这将导致循环体再次执行，从而将 i 的值从 1 增加到 2。

当 i = 2 时，循环条件的值为 False，因为位置 i 的元素大于等于 0。跳过循环主体并继续执行 if 语句。它的条件值为 True，因为 i 小于 data 的长度。因此，执行 if 部分的主体，显示 data 中的第一个正数，即 2。

5.3 其他列表操作

列表可以随着程序的运行而增长或收缩。可以在列表的任何位置插入新元素，也可以根据其值或索引删除元素。Python 还提供了一些机制，用于确定某个元素是否出现在列表中、查找列表中某个元素第一次出现时的索引、重新排列列表中的元素以及许多其他有用的任务。

向列表中插入新元素和从列表中删除元素等任务是通过向列表应用方法来执行的。与函数非常相似，方法是执行任务时可以调用的语句集合。但是，将方法应用于列表的语法与用于调用函数的语法略有不同。

把方法应用于列表时，可以使用一个语句，该语句有一个包含列表的变量[1]，后跟一个句点，然后是方法的名称。与函数调用类似，方法的名称后跟圆括号，圆括号括住了用逗号分隔的参数集合。一些方法返回一个结果。这个结果可以使用赋值语句存储在一个变量中，作为参数传递给另一个方法或函数调用，或者作为计算的一部分使用，就像函数返回的结果一样。

5.3.1 向列表中添加元素

可以通过调用 append 方法，将元素添加到现有列表的末尾。它接收一个参数，即要添加到列表中的元素。例如，考虑以下程序：

```
data = [2.71, 3.14, 1.41, 1.62]
data.append(2.30)
print(data)
```

第一行创建一个包含 4 个元素的新列表并将其存储在数据中。然后对 data 应用 append 方法，通过向列表末尾添加 2.30，将其长度从 4 增加到 5。最后，打印现在包含 2.71、3.14、1.41、1.62 和 2.30 的列表。

元素可以使用 insert 方法插入列表中的任何位置。它需要两个参数，即插入元素的索引和元素的值。当插入一个元素时，插入点右侧的任何元素的索引都要增加 1，这样新元素就有一个可用的索引。例如，下面的代码段在 data 中间插入 2.30，而不是将它附加到列表的末尾。当这个代码段执行时，它将显示 [2.71,3.14,2.30,1.41,1.62]。

```
data = [2.71, 3.14, 1.41, 1.62]
data.insert(2, 2.30)
print(data)
```

5.3.2 从列表中删除元素

pop 方法用于从列表中删除特定索引处的元素。要删除的元素

---

[1] 也可以使用相同的语法将方法应用于方括号中的列表文字，但是很少需要这样做。

的索引是作为 pop 的可选参数提供的。如果该参数被省略，那么 pop 将从列表中删除最后一个元素。pop 方法返回从列表中删除的值作为其唯一结果。当后续计算需要此值时，可以通过调用赋值语句右侧的 pop 将其存储到变量中。将 pop 应用于空列表是一个错误，就像试图从超出列表末尾的索引中删除元素一样。

也可以通过调用 remove 方法从列表中删除值。它唯一的参数是要删除的值(而不是要删除的值的索引)。当 remove 方法执行时，它从列表中删除第一个出现的参数。列表中不存在传递给 remove 的值时，将报告错误。

考虑下面的例子。它创建一个列表，然后从中删除两个元素。当第一个 print 语句执行时，它会显示[2.71,3.14]，因为从列表中删除了 1.62 和 1.41。第二个 print 语句显示 1.41，因为 1.41 是应用 pop 方法时列表中的最后一个元素。

```
data = [2.71, 3.14, 1.41, 1.62]
data.remove(1.62)    # Remove 1.62 from the list
last = data.pop()    # Remove the last element from the list

print(data)
print(last)
```

5.3.3 重新排列列表中的元素

有时，列表中包含所有正确的元素，但它们的顺序与解决特定问题的顺序不同。列表中的两个元素可使用一系列赋值语句进行交换，这些语句对列表中的各个元素进行读写操作，如下面的代码段所示。

```
# Create a list
data = [2.71, 3.14, 1.41, 1.62]

# Swap the element at index 1 with the element at index 3
temp = data[1]
data[1] = data[3]
data[3] = temp

# Display the modified list
print(data)
```

当这些语句执行时，data 初始化为[2.71,3.14,1.41,1.62]。然后，将索引 1 处的值(3.14)复制到 temp 中，接着一行代码将索引 3 处的值复制到索引 1 中。最后，将 temp 中的值复制到索引 3 处的列表中。当 print 语句执行时，它显示[2.71,1.62,1.41,3.14]。

有两种方法可以重新排列列表中的元素。reverse 方法反转列表中元素的顺序，sort 方法将元素按升序排序。reverse 和 sort 都可以应用于列表，而不需要提供任何参数[1]。

下面的示例从用户处读取一组数字，并将它们存储在一个列表中。然后它以排序的顺序显示所有值。

```
# Create a new, empty list
values = []

# Read values from the user and store them in a list until a blank line is entered
line = input("Enter a number (blank line to quit): ")
while line != "":
  num = float(line)
  values.append(num)

  line = input("Enter a number (blank line to quit): ")

# Sort the values into ascending order
values.sort()

# Display the values
for v in values:
  print(v)
```

5.3.4 搜索列表

有时，需要确定某个特定值是否出现在列表中。在其他情况下，则希望确定列表中已知值的索引。Python 的 in 操作符和 index 方法允许执行这些任务。

in 操作符用于确定一个值是否出现在列表中。要搜索的值位于

---

[1] 只有在所有元素都可以用小于操作符相互比较的情况下才能排序列表。许多 Python 类型都定义了小于操作符，包括整数、浮点数、字符串和列表等。

操作符的左侧。正在搜索的列表位于操作符的右侧。如果该值出现在列表中，则该表达式的计算结果为 True。否则计算结果为 False。

　　index 方法用于标识列表中特定值的位置。这个值作为唯一参数传递给 index。该值在列表中第一次出现时的索引将作为方法的结果返回。使用列表中不存在的参数调用 index 方法是错误的。因此，程序员有时使用 in 操作符来确定一个值是否出现在列表中，然后使用 index 方法来确定它的位置。

　　下面的例子演示了本节介绍的几个方法和操作符。它首先从用户处读取整数，并将它们存储在一个列表中。然后从用户处读取一个额外的整数。报告这个附加整数在值列表中第一次出现的位置(如果它存在的话)。如果用户输入的值列表中没有附加的整数，则显示适当的消息。

```
# Read integers from the user until a blank line is entered and store them all in data
data = []
line = input("Enter an integer (blank line to finish): ")
while line != "":
  n = int(line)
  data.append(n)

  line = input("Enter an integer (blank line to finish): ")

# Read an additional integer from the user
x = int(input("Enter one additional integer: "))

# Display the index of the first occurrence of x (if it is present in the list)
if x in data:
  print("The first", x, "is at index", data.index(x))
else:
  print(x, "is not in the list")
```

5.4　列表作为返回值和参数

　　可以从函数中返回列表。与其他类型的值一样，使用 return 关键字从函数返回列表。当 return 语句执行时，函数终止，列表返回到调用函数的位置。然后该列表可以存储在一个变量中或用于计算中。

列表也可以作为参数传递给函数。与其他类型的值一样，传递给函数的任何列表都包含在调用函数名称后的括号中。每个参数，无论是列表还是其他类型的值，都出现在函数的相应参数变量中。

包含列表的参数变量可以在函数体中使用，就像包含其他类型值的参数变量一样。但与整数、浮点数、字符串或布尔值不同，对列表参数变量所做的更改会影响传递给函数的参数以及参数变量中存储的值。特别是，使用方法(如 append、pop 或 sort)对列表进行的更改将同时更改参数变量和调用函数时提供的参数的值。

对单个列表元素执行的更新(列表的名称后跟由方括号括起来的索引，出现在赋值操作符的左侧)也修改了参数变量和调用函数时提供的参数。但对整个列表的赋值(只有列表的名称出现在赋值操作符的左边)只影响参数变量。这样的赋值不会影响调用函数时提供的参数。

列表参数和其他类型的参数之间的行为差异可能看起来是任意的，因为可以选择将一些更改同时应用于参数变量和参数，而其他更新仅更改参数变量。然而，事实并非如此。这些差异有重要的技术原因，但是这些细节超出了 Python 简介的范围。

5.5 练习

本章的所有练习都应该使用列表来解决。所编写的程序需要创建列表、修改列表、在其中定位值。有些练习还要求编写返回列表或将其作为参数的函数。

练习 110：排序

(有解答，22 行)

编写一个程序，从用户处读取整数，并将它们存储在一个列表中。程序应该继续读取值，直到用户输入 0。然后，它应该按升序显示用户输入的所有值(除了 0)，每行显示一个值。使用 sort 方法或 sorted() 函数对列表进行排序。

练习 111：倒序

(20 行)

编写一个程序，从用户处读取整数，并将它们存储在一个列表中。使用 0 作为标记值来标记输入的结束。一旦所有值都被读取，程序应该以相反的顺序显示它们(除了 0)，每行显示一个值。

练习 112：删除异常值

(有解答，44 行)

在分析作为科学实验的一部分收集的数据时，在进行其他计算之前，最好先去掉最极端的值。编写一个函数，该函数接收一个值列表和一个非负整数 n 作为参数。该函数应该创建一个新的列表副本，删除其中的 n 个最大元素和 n 个最小元素。然后它应该返回列表的新副本作为函数的唯一结果。返回列表中元素的顺序不必与原始列表中元素的顺序匹配。

编写一个 main 程序来演示函数。它应该从用户那里读取一个数字列表，并通过调用前面描述的函数，从该列表中删除两个最大和两个最小的值。显示删除了异常值的列表，然后显示原始列表。如果用户输入的值小于 4，则程序应该生成适当的错误消息。

练习 113：避免重复

(有解答，21 行)

本练习将创建一个程序，从用户处读取单词，直到用户输入空行。在用户输入空白行之后，程序应该显示一次用户输入的每个单词。单词应该按照它们最初输入的顺序显示。例如，如果用户输入：

```
first
second
first
third
second
```

程序就应该显示：

```
first
second
third
```

练习 114：负数、零和正数

(有解答，36 行)

创建一个程序，从用户处读取整数，直到输入空行。一旦所有整数都被读取，程序应该显示所有负数，然后是所有的零，最后是所有正数。在每个组中，数字应该按照用户输入的顺序显示。例如，如果用户输入值 3、-4、1、0、-1、0 和 -2，则程序应该输出值 -4、-1、-2、0、0、3 和 1。程序应该将每个值显示在单独一行上。

练习 115：正因数表

(36 行)

正整数 n 的正因数是一个小于 n、且能整除 n 的正整数。编写一个函数来计算正整数的所有正因数。该整数作为函数的唯一参数传递。该函数返回一个包含所有正因数的列表，作为其唯一结果。通过编写 main 程序来完成这个练习，main 程序通过从用户处读取一个值，并显示其正因数列表来演示该函数。确保 main 程序只在解答没有导入另一个文件时运行。

练习 116：完全数

(有解答，35 行)

当整数 n 的正因数的总和等于 n 时，n 就是完全数。例如，28 是一个完全数，因为其正因数是 1、2、4、7 和 14，1 + 2 + 4 + 7 + 14 = 28。

编写一个函数，来确定某正整数是否为完全数。函数接收一个参数。如果参数是完全数，函数将返回 True，否则返回 False。另外，编写一个 main 程序，使用函数来识别和显示 1 到 10 000 之间的所有完全数。完成此任务时，将该解答导入到练习 115。

练习 117：只看单词

(38 行)

本练习将创建一个程序，来识别用户输入的字符串中的所有单词。首先编写一个函数，该函数将字符串作为唯一的参数。函数应该返回字符串中的单词列表，并删除单词边缘的标点符号。必须考虑的标点符号包括逗号、句号、问号、连字符、撇号、感叹号、冒号和分号。不要删除出现在单词中间的标点符号，例如构成缩写的撇号。举个例子，如果给函数提供了字符串"Contractions include: don't, isn't, and wouldn't"，那么函数应该返回列表["Contractions", "include", "don't ", "isn't ", "and", "wouldn't "]。

编写一个 main 程序来演示函数。它应该从用户那里读取一个字符串，然后显示字符串中的所有单词，去掉标点符号。完成练习 118 和 167 时，需要导入该解答。因此，应该确保 main 程序只在文件没有导入另一个程序时运行。

练习 118：逐字回文

(34 行)

之前的练习 75 和 76 介绍了回文的概念。回文只考虑了字符串中的字符，忽略了空格和标点符号，这个字符串中的字符无论向前还是向后看都是相同的。回文通常是一个字符一个字符地考虑，但回文的概念可以扩展到更大的单位。举个例子，句子"Is it crazy how saying sentences backwards creates backwards sentences saying how crazy it is?"不是逐字符的回文，而是逐字的回文(忽略大小写和标点符号)。其他逐字回文的例子包括"Herb the sage eats sage, the Herb"和"Information school graduate seek graduate school Information"。

创建一个程序，从用户处读取字符串。程序应该报告输入的字符串是不是逐字回文。在确定结果时忽略空格和标点符号。

练习 119：低于和高于平均水平

(44 行)

编写一个程序，从用户处读取数字，直到用户输入空行。程序应该显示用户输入的所有值的平均值。然后程序应该显示所有低于平均值的值，然后显示所有平均值(如果有)，最后显示所有高于平均值的值。在每个值列表之前应该显示一个适当标签。

练习 120：格式化列表

(有解答，41 行)

用英语编写一个项列表时，通常用逗号分隔各个项。此外，and 一词通常包括在最后一项之前，除非列表只包含一项。考虑以下四个列表：

```
apples
apples and oranges
apples, oranges and bananas
apples, oranges, bananas and lemonss
```

编写一个函数，将字符串列表作为它的唯一参数。函数应该返回一个字符串，该字符串包含列表中的所有项，按照前面描述的方式进行格式化，作为它的唯一结果。虽然在前面展示的示例中，列表只包含四个或更少的元素，但是对于任何长度的列表，函数的行为都应该是正确的。包含一个 main 程序，它从用户处读取几个项，

第 5 章 列　　表

通过调用函数对它们进行格式化，然后显示函数返回的结果。

练习 121：随机的彩票号码

(有解答，28 行)

为赢得某项彩票的头奖，必须将彩票上的所有 6 个数字与彩票组织者抽到的 1~49 之间的 6 个数字匹配。编写一个程序，生成随机选择 6 个数字的彩票。确保所选的 6 个数字不包含任何重复。按升序显示数字。

练习 122：Pig Latin

(32 行)

Pig Latin 是通过转化英语单词而构成的一种语言。虽然这种语言的来源是未知的，但它至少在 19 世纪的两份文件中被提及，表明它已经存在了 100 多年。将英语翻译成 Pig Latin 的规则如下：

- 如果单词以辅音字母开头(包括 y)，那么单词开头的所有字母，直到第一个元音字母(不包括 y)，都将被删除，然后添加到单词末尾，后面跟着 ay。例如，computer 变成 omputercay，think 变成 inkthay。
- 如果单词以元音开头(不包括 y)，那么把 way 加到单词的末尾。例如，algorithm 变成 algorithmway，office 变成 officeway。

编写一个程序，从用户那里读取一行文本。然后，程序应该将该行翻译成 Pig Latin 并显示结果。可以假设用户输入的字符串只包含小写字母和空格。

练习 123：Pig Latin 进步了

(51 行)

扩展练习 122 的解答，以便正确处理大写字母和标点符号，如逗号、句号、问号和感叹号。如果一个英语单词以大写字母开头，那么它的 Pig Latin 表示形式也应该以大写字母开头,并将大写字母移到末尾,这个词应该改成小写。例如,Computer 应该变成 Omputercay。如果一个单词以标点符号结束,那么在转换完成后,标点符号应该仍在单词的末尾。例如 Science!应该成为 Iencescay!。

练习 124:最佳拟合线

(41 行)

最佳拟合线是最接近 n 个数据点集合的直线。在这个练习中,假设集合中的每个点都有一个 x 坐标和一个 y 坐标。符号 $\bar{x}$ 和 $\bar{y}$ 分别表示集合中的 x 平均值和 y 平均值。最佳拟合直线由方程 $y = mx + b$ 表示,其中 m 和 b 采用以下公式计算:

$$m = \frac{\sum xy - \frac{(\sum x)(\sum y)}{n}}{\sum x^2 - \frac{(\sum x)^2}{n}}$$

$$b = \bar{y} - m\bar{x}$$

编写一个程序,从用户处读取点的集合。用户在单独一行上输入第一个 x 坐标,然后在另一行上输入第一个 y 坐标。允许用户继续输入坐标,在单独的行上输入 x 和 y 值,直到程序为 x 坐标读取空行为止。以 $y = mx + b$ 的形式显示最佳拟合线的公式,用前面公式计算的值替换 m 和 b。例如,如果用户输入坐标(1, 1)、(2, 2.1)和(3, 2.9),则程序应该显示 $y = 0.95x + 0.1$。

练习 125:洗牌

(有解答,49 行)

一副标准的扑克牌包含 52 张牌。每张牌有四种花色中的一种,并带有一个值。花色通常是黑桃、红心、方块和梅花,而值是 2 到 10、Jack、Queen、King 和 Ace。

每张纸牌可用两个字符表示。第一个字符是纸牌的面值,值 2

到 9 直接表示。字符 T、J、Q、K 和 A 分别代表 10、Jack、Queen、King 和 Ace 的值。第二个字符用于表示纸牌的花色。它通常是小写字母：黑桃是 s，红心是 h，方块是 d，梅花是 c。下表提供了一些纸牌及其双字符表示的示例。

牌	缩写
黑桃 j	Js
梅花 2	2c
方块 10	Td
红心 A	Ah
黑桃 9	9s

首先编写一个名为 createDeck 的函数。它使用循环来创建一副完整的纸牌，方法是将 52 张纸牌的两字符缩写存储到一个列表中。返回卡片列表作为函数的唯一结果。函数不需要任何参数。

编写第二个名为 shuffle 的函数，用于随机化列表中纸牌的顺序。可以用来洗牌的一种技术是访问列表中的每个元素，并将其与列表中的另一个随机元素交换。必须编写自己的洗牌循环。不能使用 Python 的内置 shuffle() 函数。

使用前几段中描述的两个函数创建一个 main 程序，该 main 程序显示洗牌前后的一副牌。确保 main 程序只在函数没有导入另一个文件时运行。

一个好的洗牌算法是无偏的，这意味着当算法完成时，每个元素的不同排列都有相同的可能性。虽然前面在这个问题中描述的方法建议，按顺序访问每个元素，并与一个随机索引的元素交换，但是这种算法是有偏的。特别是，在原始列表中稍后出现的元素更可能出现在打乱列表的后面。与直觉相反，通过按顺序访问每个元素，并将其交换到当前元素的位置和列表末尾之间的一个随机索引，而不是随机选择任何索引，可以实现无偏洗牌。

练习 126：发牌

(44 行)

在许多纸牌游戏中，每个玩家在洗完牌后都会得到特定数量的牌。编写一个函数 deal()，它以手牌的数量、每手牌的数量和一副牌作为它的三个参数。该函数应该返回一个包含已发所有手牌的列表。每手牌都表示为一个纸牌列表。

当处理手牌时，函数应该修改作为参数传递给它的牌组，当它发到玩家的手上时，将每一张牌从牌组中移除。发牌时，习惯上先给每个玩家发一张牌，再给玩家发下一张牌。函数在构建玩家的牌时，应该遵循这一惯例。

使用练习 125 的解答帮助构建一个 main 程序，这个 main 程序创建一副牌并洗牌，然后给四个玩家发五张牌。显示所有玩家的牌，以及牌发完后牌组中剩余的牌。

练习 127：列表已经排好序了吗？

(41 行)

编写一个函数，来确定值列表是否按顺序排序(升序或降序)。如果列表已经排序，函数应该返回 True，否则应该返回 False。编写一个 main 程序，从用户那里读取一个数字列表，然后使用函数报告这个列表是否已排序。

> 在完成这个练习时，一定要考虑以下问题：列表是否为空？包含一个元素的列表如何处理？

练习 128：确定元素个数

(有解答，48 行)

Python 的标准库包含一个名为 count()的方法，它可以确定特定值在列表中出现的次数。这个练习创建一个名为 countRange()的新函数。它确定并返回列表中大于或等于某个最小值而小于某个最大值的元素数。函数接收三个参数：列表、最小值和最大值。它返回一个大于或等于的整数结果。

包括一个 main 程序，为几个不同的列表，最小值和最大值演示函数。确保程序对整数列表和浮点数列表都能正确工作。

练习 129：标记字符串

(有解答，47 行)

标记化是将字符串转换为子字符串列表的过程，称为标记。许多情况下，使用标记列表要比使用原始字符串容易得多，因为原始字符串可能具有不规则的间距。某些情况下，还需要进行大量的工作来确定一个标记在何处结束，下一个标记在何处开始。

在数学表达式中，标记是操作符、数字和括号等项。在这个练习及后续练习中考虑的运算符是 *、/、^、+。操作符和括号很容易识别，因为标记始终是单个字符，而该字符从不是另一个令牌的一部分。识别数字稍微有点困难，因为标记可能包含多个字符。任何连续的数字序列都应该被视为一个数字标记。

编写一个函数，该函数将一个包含数学表达式的字符串作为其唯一参数，并将其分解为一个标记列表。每个标记应该是括号、操作符或数字。为简单起见，这个练习只处理整数。返回标记列表作为函数的唯一结果。

可以假设，传递给函数的字符串总是包含一个有效的数学表达式，该表达式由括号、操作符和整数组成。但是，函数必须处理这些元素之间的可变空白(包括没有空白)。包含一个 main 程序，通过从用户处读取表达式并打印标记列表来演示标记函数。确保在将包含解答的文件导入另一个程序时，main 程序不会运行。

练习 130：一元运算符和二元运算符

(有解答，45 行)

有些数学运算符是一元的，有些是二元的。一元操作符作用于一个值，二元操作符作用于两个值。例如，在表达式 2 *(-3)中，*是一个二元运算符，因为它同时作用于 2 和-3，而-是一元运算符，因为它只作用于 3。

操作符的符号并不总是足以确定它是一元的还是二元的。例如，在前面的表达式中，-运算符是一元的，而在诸如 2 - 3 的表达式中，相同的字符"-"用于表示二进制-运算符。在对表示数学表达式的标记列表执行其他有趣的操作之前，必须消除+运算符也存在的这种歧义。

创建一个函数，在标记列表中标识一元运算符+和-运算符，并分别用 u+和 u-替换它们。函数将一个数学表达式的标记列表作为它的唯一参数，并返回一个新的标记列表，其中被替换的一元+和-运算符是它的唯一结果。如果+或-运算符是列表中的第一个标记，或者紧挨着它前面的标记是运算符或左括号，那么它就是一元的。否则操作符就是二元的。

编写一个 main 程序，通过读取和标记用户输入的表达式中的一元运算符来演示函数的正确工作。将函数导入另一个程序时，main 程序不应执行。

练习 131：中缀到后缀

(63 行)

数学表达式通常以中缀形式编写，运算符出现在它们作用的操作数之间。虽然这是一种常见的形式，但也可以用后缀形式表示数学表达式，即运算符出现在其所有操作数之后。例如，中缀表达式 3 + 4 以后缀形式写成 3 4 +。可以使用以下算法将中缀表达式转换为后缀形式：

创建一个新的空列表 operators
创建一个新的空列表 postfix

for 中缀表达式中的每个标记

if 标记是整数，then
　　将标记追加到 postfix
if 标记是操作符，then
　　while　operators 不是空的 and
　　　　operators 中的最后一项不是开括号 and
　　　　precedence(标记)<precedence(operators 中的最后一项)　do
　　　　从 operators 中删除最后一项并将其追加到 postfix
　　将标记追加到 operators
if 标记是一个左括号，then
　　将标记追加到 operators
if 标记是右括号，then
　　while　operators 中的最后一项不是左括号 do
　　　　从 operators 中删除最后一项并将其追加到 postfix
　　从 operators 中移除开括号

while　operators 不是空列表 do
　　从 operators 中删除最后一项并将其追加到 postfix

return postfix 作为算法的结果

请用练习 129 和 130 的答案标记一个数学表达式，并找出其中的一元运算符。然后使用上面的算法将表达式从中缀形式转换为后缀形式。实现上述算法的代码应该放在一个函数中，该函数将表示中缀表达式(标记了一元运算符)的标记列表作为其唯一参数。它应该返回一个表示等效后缀表达式的标记列表作为其唯一结果。包含一个 main 程序，读取用户输入的中缀形式的表达式，并以后缀形式显示它，来演示中缀到后缀函数。

阅读练习 132 时，从中缀形式转换为后缀形式的目的变得很明显。完成这道题时，会发现练习 96 和 97 的答案很有帮助。应该能够不做任何修改地使用练习 96 的解答，但是需要扩展练习 97 的解答,以便为一元运算符返回正确的优先级。一元运算符的优先级应该高于乘法和除法运算符,但低于指数运算符。

练习 132:计算后缀形式的表达式

(63 行)

求后缀表达式比求中缀表达式更简单,因为它不包含任何括号,也没有要考虑的运算符优先规则。后缀表达式可以用以下算法求值:

> 创建一个新的空列表 values
>
> for 后缀表达式中的每个标记
>   if 标记是数字,then
>   将其转换为整数并将其追加到 values 中
>   else if 标记为一元减号 then
>   从 values 的末尾删除项
>   使项变号,并将变号的结果追加到 values 上
>   else if 标记是二元操作符,then
>   从 values 的末尾删除一项,并将其称为 right。
>   从 values 的末尾删除一项,并将其称为 left
>   计算将运算符应用于 left、right 的结果
>   将结果追加到 values 上
>
> return values 中的第一项作为表达式的值

编写一个程序,从用户那里读取中缀形式的数学表达式,将其转换为后缀形式,计算并显示其值。请用练习 129、130 和 131 的答案以及上面所示的算法来解决这个问题。

练习 131 和 132 中提供的算法不执行任何错误检查。结果,如果提供了无效输入,程序可能崩溃或产生错误结果。这些练习的算

法可以扩展到检测无效输入，并对其作出合理的响应。这些留给有兴趣的学生作为独立的练习。

练习 133：列表是否包含子列表？

(44 行)

子列表是更大列表的一部分。子列表可以是包含单个元素、多个元素甚至根本没有元素的列表。例如，[1]、[2]、[3]和[4]都是[1, 2, 3, 4]的子列表。该列表[2, 3]也是[1, 2, 3, 4]的一个子列表，但[2, 4]不是[1, 2, 3, 4]的子列表，因为在较长的列表中，元素 2 和 4 不相邻。空列表是任何列表的子列表。因此，[]是[1, 2, 3, 4]的子列表。列表是其本身的一个子列表，这意味着[1, 2, 3, 4]也是[1, 2, 3, 4]的子列表。

这个练习将创建一个函数 isSublist()，它决定一个列表是不是另一个列表的子列表。函数应该使用两个列表 larger 和 smaller 作为其参数。当且仅当 smaller 是 larger 的子列表时，函数应该返回 True。编写一个 main 程序来演示函数。

练习 134：生成列表的所有子列表

(有解答，41 行)

使用练习 133 中子列表的定义，编写一个函数，该函数返回一个列表，它包含某列表中所有可能的子列表。例如[1, 2, 3]的子列表有[], [1], [2], [3], [1, 2], [2, 3]和[1, 2, 3]。注意，函数返回的列表总是至少包含一个空列表，因为空列表是每个列表的子列表。包含一个 main 程序，通过显示几个不同列表的所有子列表来演示函数。

练习 135：埃拉托色尼筛法

(有解答，33 行)

埃拉托色尼筛法是 2000 多年前发展起来的一种技术，可以很容易地找到 2 到某个极限之间的所有质数，比如 100。算法描述如下：

  写下从 0 到极限的所有数
  划掉 0 和 1，因为它们不是质数

  令 p = 2
  while   p 小于极限 do
    划掉所有 p 的倍数(但不包括 p 本身)
    令 p 等于列表中未划去的下一个数字

  将所有未划去的数字都报告为质数

  这个算法的关键是，在一张纸上划掉第 n 个数字是相对容易的。对于计算机来说，这也是一个简单的任务——向 range() 函数提供第三个参数时，for 循环可以模拟这种行为。一个数字被划掉时，我们知道它不再是质数，但它仍然占据纸上的空间，在计算以后的质数时仍然必须考虑它。因此，不应该通过从列表中删除一个数字来模拟删除该数字。相反，应该模拟用 0 替换一个数字来删除它。然后，一旦算法完成，列表中的所有非零值都是质数。

  创建一个 Python 程序，该程序使用该算法来显示 2 到用户输入的极限之间的所有质数。如果正确地实现了算法，应该能够在几秒钟内显示小于 1 000 000 的所有素数。

> 这种寻找质数的算法并不是埃拉托色尼唯一出名的地方。他其他值得注意的成就包括计算地球的周长和地轴的倾斜度。他还担任过亚历山大图书馆的馆长。

第6章 字　　典

　　列表和字典之间有许多相似之处。与列表一样，字典允许在一个变量中存储多个甚至许多值。列表中的每个元素都有唯一的整数索引，这些索引必须是从 0 开始依次递增的整数。类似地，字典中的每个值都有一个与之关联的唯一键，但是字典的键比列表的索引更灵活。字典的键可以是整数，也可以是浮点数或字符串。当键是数字时，不必从零开始，也不必是连续的。当键是字符串时，可以是任何字符组合，包括空字符串。字典中的所有键必须是不同的，就像列表中的所有索引是不同的一样。

　　字典中的每个键都必须有一个与之相关联的值。与键关联的值可以是整数、浮点数、字符串或布尔值。它也可以是一个列表，甚至是另一个字典。字典键及其对应的值通常被称为键-值对。虽然字典中的键必须是不同的，但对值没有这个限制。因此，同一个值可以与多个键关联。

　　从 Python 3.7 开始，字典中的键值对总是按照它们添加到字典中的顺序存储[1]。每当在字典添加中一个新的键值对时，它就添加到现有集合的末尾。没有将键值对插入现有字典中间的机制。从字典中删除键-值对不会改变字典中其余键-值对的顺序。

保存字典的变量是使用赋值语句创建的。空字典不包含任何键值对，由{}表示(一个左花括号紧接一个右花括号)。在花括号中包含用逗号分隔的键-值对集合，可以创建非空字典。冒号用于在每个键-值对中将键与其值分隔开。例如，下面的程序用三个键-值对创建一个字典，其中键是字符串，值是浮点数。每个键值对都将一个常见数学常数的名称与其值关联起来。然后通过调用 print() 函数来显示所有的键值对。

```
constants = {"pi": 3.14, "e": 2.71, "root 2": 1.41}
print(constants)
```

6.1 访问、修改和添加值

在字典中访问值类似于在列表中访问值。当列表中的值的索引已知时，可以使用列表的名称和方括号中的索引来访问该位置的值。类似地，当字典中与值关联的键已知时，可以使用字典的名称和方括号中的键来访问与该键关联的值。

修改字典中的现有值和向字典中添加新的键-值对都是使用赋值语句执行的。字典的名称和方括号中的键放在赋值操作符的左边，与键关联的值放在赋值操作符的右边。如果键已经存在于字典中，那么赋值语句将用赋值操作符右边的值替换键的当前值。如果该键在字典中尚未出现，则向其添加一个新的键值对。下面的程序演示了这些操作。

```
# Create a new dictionary with 2 key-value pairs
```

---

[1] 在 Python 早期版本中，不能保证键值对的存储顺序就是它们添加到字典中的顺序。

```
results = {"pass": 0, "fail": 0}

# Add a new key-value pair to the dictionary
results["withdrawal"] = 1

# Update two values in the dictionary
results["pass"] = 3
results["fail"] = results["fail"] + 1

# Display the values associated with fail, pass and withdrawal respectively
print(results["fail"])
print(results["pass"])
print(results["withdrawal"])
```

当这个程序执行时,它创建了一个名为 results 的字典,该字典最初有两个键 pass 和 fail。与每个键关联的值是 0。第三个键是 withdraw,它使用赋值语句将值 1 添加到字典中。然后使用第二个赋值语句将与 pass 关联的值更新为 3。接下来的代码行读取与 fail 相关的当前值,即 0,然后向它加 1,将这个新值存储回字典中,替换以前的值。当打印值 1(当前与 fail 相关的值)显示在第一行时,3(当前与 pass 相关的值)显示在第二行,1(当前与 withdrawal 相关的值)显示在第三行。

6.2 删除键值对

使用 pop 方法从字典中删除键值对。调用方法时必须提供一个参数,即要删除的键。当该方法执行时,它将从字典中删除键和与之关联的值。与列表不同,通过调用 pop 而不带任何参数,不可能从字典中删除最后一个键-值对。

pop 方法返回从字典中删除的键所关联的值。此值可以使用赋值语句存储到变量中,也可以用于需要值的任何地方,例如将其作为参数传递给另一个函数或方法调用,或者作为算术表达式的一部分。

6.3 其他字典操作

一些程序将键值对添加到从用户那里读取键或值的字典中。一旦所有的键值对都存储在字典中，就有必要确定有多少个键，某个特定的键在字典中是否存在，或者某个特定的值在字典中是否存在。Python 提供了函数、方法和运算符，来实现这些任务。

len()函数以前用于确定列表中元素的数量，现在也可以用于确定字典中有多少个键值对。字典作为函数的唯一参数传递，键-值对的数量作为函数的结果返回。如果作为参数传递的字典为空，则 len()函数返回 0。

in 操作符可用于确定字典中是否存在特定的键或值。在搜索键时，键出现在 in 操作符的左侧，而字典则出现在其右侧。如果键存在于字典中，则运算符的计算结果为 True。否则计算结果为 False。in 操作符返回的结果可以在任何需要布尔值的地方使用，包括在 if 语句或 while 循环的条件中。

in 操作符与 values()方法一起使用，可以确定一个值是否出现在字典中。要搜索的值出现在 in 操作符的左侧，而应用了 values 方法的 dictionary 则出现在其右侧。例如，下面的代码段确定字典 d 中是否有等于当前变量 x 存储的值。

```
if x in d.values():
  print("At least one of the values in d is", x)
else:
  print("None of the values in d are", x)
```

6.4 循环和字典

for 循环可用于遍历字典中的所有键，如下所示。每次循环体执行时，都会将字典中的不同键存储到 for 循环的变量 k 中。

第6章 字　　典

```
# Create a dictionary
constants = {"pi": 3.14, "e": 2.71, "root 2": 1.41}

# Print all of the keys and values with nice formatting
for k in constants:
  print("The value associated with", k, "is", constants[k])
```

当这个程序执行时，首先创建一个包含三个键-值对的新字典。然后，for 循环遍历字典中的键。字典中的第一个键 pi 存储在 k 中，循环体执行。它打印出一个有意义的消息，其中包括 pi 和它的值，即 3.14。然后，控制流返回到循环顶部，e 存储到 k 中。循环体第二次执行，并显示一条消息，指示 e 的值为 2.71。最后，循环执行第三次，k 等于 root 2，并显示最后的消息。

还可以使用 for 循环遍历字典中的值(而不是键)。为此，应将不带参数的 values()方法应用于字典，来创建 for 循环使用的值集合。例如，下面的程序计算字典中所有值的和。当它执行时，constants.values()将是一个包含 3.14、2.71 和 1.41 的集合。当 for 循环运行时，这些值中的每一个都存储在 v 中，这样就可以在不使用任何字典键的情况下计算总和。

```
# Create a dictionary
constants = {"pi": 3.14, "e": 2.71, "root 2": 1.41}

# Compute the sum of all the value values in the dictionary
total = 0
for v in constants.values():
  total = total + v

# Display the total
print("The total is", total)
```

与 for 循环相比，使用 while 循环可更好地解决一些涉及字典的问题。例如，下面的程序使用 while 循环读取用户输入的字符串，直到用户输入 5 个唯一值为止。然后显示所有字符串及其计数。

```
# Count how many times each string is entered by the user
counts = {}

# Loop until 5 distinct strings have been entered
```

```
while len(counts) < 5:
  s = input("Enter a string: ")

  # If s is already a key in the dictionary then increase its count by 1. Otherwise
      add s to the
  # dictionary with a count of 1.
  if s in counts:
    counts[s] = counts[s] + 1
  else:
    counts[s] = 1

# Displays all of the strings and their counts
for k in counts:
  print(k, "occurred", counts[k], "times")
```

当这个程序执行时，首先创建一个空字典。然后计算 while 循环条件。它使用 len()函数确定字典中有多少个键值对。由于键值对的数量最初为 0，因此条件的计算结果为 True，循环体执行。

每次循环体执行时，都会从用户处读取一个字符串。然后使用 in 运算符判断字符串是否已经是字典中的键。如果是，则与键关联的计数增加 1。否则，该字符串将作为值为 1 的新键添加到字典中。循环继续执行，直到字典包含 5 个键值对。一旦出现这种情况，用户输入的所有字符串以及相关的值都会显示出来。

6.5 字典作为参数和返回值

字典可以作为参数传递给函数，就像其他类型的值一样。与列表一样，对包含字典的参数变量进行更改，可以修改参数变量和传递给函数的参数。例如，插入或删除键值对将同时修改参数变量和参数，就像使用赋值语句修改与字典中的键相关联的值一样。但对整个字典的赋值(只有字典的名称出现在赋值操作符的左边)只影响参数变量。它不修改传递给函数的参数。与其他类型一样，字典是使用 return 关键字从函数返回的。

6.6 练习

虽然本章的许多练习可以用列表或 if 语句来解决,但大多数(甚至所有)练习都有非常适合于字典的解。因此,应该使用字典来解答所有这些练习,而不是(或额外)使用前面介绍的 Python 特性。

练习 136:反向查找

(有解答,45 行)

编写一个名为 reverseLookup() 的函数,查找字典中映射到特定值的所有键。该函数将字典和要搜索的值作为其唯一的参数,从字典中返回一个(可能是空的)键列表,这些键映射到提供的值。

包含演示 reverseLookup() 函数的 main 程序,作为本练习的解答的一部分。程序应该创建一个字典,然后在返回多个键、单个键和无键时显示 reverseLookup() 函数工作正常。确保 main 程序只在包含此练习解答的文件未导入其他程序时运行。

练习 137:两个骰子的模拟

(有解答,43 行)

这个练习将模拟两个骰子滚动 1000 次。首先编写一个函数,模拟滚动一对六面骰子。函数不接收任何参数,返回两个骰子的滚动次数作为它的唯一结果。

编写一个 main 程序,使用该函数来模拟滚动两个六面骰子 1000 次。当程序运行时,它应该计算每个总数发生的次数。然后,它应该显示一个汇总这些数据的表。将每个总数的频率表示为所执行的滚动次数的百分比。程序还应该显示概率理论对每个总数的期望百分比。示例输出如下所示。

```
Total      Simulated       Expected
Percent    Percent
2                2.90          2.78
3                6.90          5.56
```

4	9.40	8.33
5	11.90	11.11
6	14.20	13.89
7	14.20	16.67
8	15.00	13.89
9	10.50	11.11
10	7.90	8.33
11	4.50	5.56
12	2.60	2.78

练习 138：发短信

(21 行)

在一些较旧的手机上，可以用数字键盘发送短信。因为每个键都有多个相关联的字母，所以大多数字母都需要多次按键。按一次数字将生成该键列出的第一个字符。按数字 2、3、4 或 5 次会产生第二个、第三个、第四个或第五个字符。

键	符号
1	.,?!:
2	A B C
3	D E F
4	G H I
5	J K L
6	M N O
7	P Q R S
8	T U V
9	W X Y Z
0	空格

编写一个程序，显示用户输入消息所需的按键。构造一个字典，从每个字母或符号映射到生成它所需的按键。然后使用字典创建和显示用户消息所需的按键。例如，如果用户输入 Hello, World!，程序就应该输出 4433555555666110966677755531111。确保程序同时

处理大小写字母。忽略上表中没有列出的字符，如分号和括号。

练习139：莫尔斯电码

(15 行)

莫尔斯电码是一种使用短横线和圆点来表示数字和字母的编码方案。本练习要求编写一个程序，使用字典来存储从这些符号到莫尔斯电码的映射。用句号表示点，用连字符表示短横线。从字符到短横线和点的映射如表 6.1 所示。

表 6.1　字母和数字的莫尔斯电码

字符	代码	字符	代码	字符	代码	字符	代码
A	.-	J	.---	S	...	1	.----
B	-...	K	-.-	T	-	2	..---
C	-.-.	L	.-..	U	..-	3	...--
D	-..	M	--	V	...-	4	....-
E	.	N	-.	W	.--	5	.....
F	..-.	O	---	X	-..-	6	-....
G	--.	P	.--.	Y	-.--	7	--...
H	....	Q	--.-	Z	--..	8	---..
I	..	R	.-.	0	-----	9	----.

程序应该读取来自用户的消息。然后，它应该将信息中的所有字母和数字转换成莫尔斯电码，在每个短横线和圆点序列之间留出空间。程序应该忽略上表中没有列出的任何字符。Hello, World!的莫尔斯电码如下所示：

　　　　.... . .-.. .-.. --- .-- --- .-. .-.. -..

> 莫尔斯电码最初是在19世纪发展起来的，用于电报线路。在160多年后的今天，人们仍然在使用它。

练习 140：邮政编码

(24 行)

加拿大邮政编码的第一、三和五个字符是字母，第二、四和六个字符是数字。地址所在的省或地区可以根据邮政编码的第一个字符来确定，如下表所示。目前有效的邮政编码不以 D、F、I、O、Q、U、W 或 Z 开头。

省/地区	第一字符
纽芬兰	A
新斯科舍	B
爱德华王子岛	C
新不伦瑞克	E
魁北克	G, H 和 J
安大略省	K, L, M, N 和 P
马尼托巴省	R
萨斯喀彻温省	S
阿尔伯塔省	T
不列颠哥伦比亚省	V
努勒维特	X
西北地区	X
育空	Y

邮政编码中的第二个字符标识地址是农村还是城市。如果该字符为 0，则该地址为农村地址。否则就是城市。

创建一个程序，从用户处读取邮政编码，并显示与之相关的省或地区，以及地址是城市还是农村。例如，如果用户输入 T2N 1N4，那么程序应该指出邮政编码是阿尔伯塔省的城市地址。如果用户输入 X0A 1B2，那么程序应该指出邮政编码是努勒维特或西北地区的农村地址。使用字典将邮政编码的第一个字符映射到省或地区名称。如果邮政编码以无效字符开头，或者邮政编码中的第二个字符不是数字，则显示有意义的错误消息。

练习 141：用英语写数字

(65 行)

尽管支票作为一种支付方式的普及程度近年来有所下降，但一些公司仍向员工或供应商发放支票。支付的金额通常出现在支票上两次，一次用数字，另一次用英语单词。以两种不同的形式重复支付支票金额，会使无良的雇员或供应商在存入支票前更难修改支票金额。

本练习的任务是创建一个函数，该函数仅接收 0 到 999 之间的整数作为参数，并返回一个包含该数字的英文单词的字符串。例如，如果函数的参数是 142，那么函数应该返回"one hundred forty two"。使用一个或多个字典来实现解决方案，而不是使用大型的 if/elif/else 结构。包含一个 main 程序，它从用户那里读取一个整数，并以英文单词显示其值。

练习 142：独特的字符

(有解答，16 行)

创建一个程序，确定并显示用户输入的字符串中不同字符的数量。例如，Hello, World! 有 10 个不同的字符，而 zzz 只有一个相同的字符。用字典或集来解决这个问题。

练习 143：字谜

(有解答，39 行)

如果两个单词包含所有相同的字母，但顺序不同，它们就是字谜。例如，"evil"和"live"是字谜，因为它们都包含 e、i、l 和 v。创建一个程序，从用户那里读取两个字符串，判断它们是不是字谜，并报告结果。

练习 144：再来一次字谜游戏

(48 行)

字谜的概念可以扩展到多个单词。例如，如果忽略字母大小写和字母间距，"William Shakespeare"和"I am a weakish speller"是字谜。

扩展练习 143 的程序，使之能检查两个短语是否是字谜。在进行判断时，程序应该忽略大小写、标点符号和空格。

练习 145：Scrabble 评分

(有解答，18 行)

在 Scrabble 游戏中，每个字母都有与之相关的分数。一个单词的总分是其字母得分的总和。更常见的字母的分数更低，而不太常见的字母的分数更高。每个字母的相关分数如下：

分数	字母
1	A、E、I、L、N、O、R、S、T 和 U
2	D 和 G
3	B、C、M 和 P
4	F、H、V、W 和 Y
5	K
8	J 和 X
10	Q 和 Z

编写一个程序，计算并显示单词的 Scrabble 数。创建一个从字母映射到分值的字典。然后用字典来计算分数。

> Scrabble 包含一些乘以字母值或整个单词值的方块。在这个练习中，忽略了这些方块。

第 6 章 字　典

练习 146：制作一张宾果卡

(有解答，58 行)

一张宾果卡(Bingo Card)由 5 列 5 个数字组成，分别用字母 B、I、N、G 和 O 标记。每个字母下有 15 个数字。具体而言，可以出现在 B 范围内的数字从 1 到 15，可以出现在 I 范围内的数字从 16 到 30，可以出现在 N 范围内的数字从 31 到 45，等等。

编写一个函数，创建一个随机宾果卡，并将它存储在一个字典中。键是字母 B、I、N、G 和 O。值是出现在每个字母下面的五个数字的列表。编写第二个函数来显示宾果卡，并适当地标记列。使用这些函数来编写一个程序，显示一个随机的宾果卡。确保 main 程序只在包含解答的文件未导入其他程序时运行。

> 注意，宾果卡的中间通常有一个"空闲"空间。这个练习不考虑该空间。

练习 147：检查获胜的卡片

(102 行)

一张获胜的宾果卡包含一行 5 个数字，它们都已被调用。玩家通常通过划掉或用涂抹的宾果来记录被调用的数字。本练习将通过在宾果卡字典中用 a0 替换一个数字来标记该数字已被调用。

编写一个函数，以表示宾果卡的字典作为其唯一参数。如果该卡包含一行 5 个 0(垂直、水平或对角线)，则函数应该返回 True，表示该卡已经赢了。否则函数返回 False。

创建一个 main 程序来演示函数，方法是创建几个宾果卡片，显示它们，并指示它们是否包含获胜行。演示函数时，应该使用至少一张画有水平线的宾果卡，至少一张画有垂直线的宾果卡，至少一张画有对角线的宾果卡，以及至少一张划掉一些数字但不包含获胜行的宾果卡。在完成这个练习时，可能希望将其解答导入练习 146。

提示：因为宾果卡上没有负数，所以找到一行 5 个 0 就等于找到 5 个元素之和为 0 的一行。求和问题比较容易解决。

练习 148：玩宾果游戏

(88 行)

这个练习将编写一个程序来模拟一张卡片的宾果游戏。首先生成所有有效 Bingo 调用(B1 到 O75)的列表。创建列表后，可以通过调用 random 模块中的 shuffle()函数来随机化其元素的顺序。然后，程序应该使用列表外的调用，并划掉卡片上的数字，直到卡片包含获胜行。模拟 1000 场游戏，并报告在卡片获胜之前必须进行的最小、最大和平均调用数。完成这个练习时，将答案导入练习 146 和 147 是很有帮助的。

第7章 文件和异常

前面创建的程序都从键盘读取输入。因此，必须在每次程序运行时重新输入所有输入值。这是低效的，特别是对于需要大量输入的程序。类似地，程序在屏幕上显示了所有结果。当只打印几行输出时，这种方法可以很好地工作，但是如果输出的结果太多，以至于无法从屏幕上读取，或者输出需要其他程序进行进一步分析，那么这种方法就不太实用。编写有效使用文件的程序将允许解决这些问题。

文件是相对永久的。存储在其中的值在程序完成后和计算机关闭时会保留下来。所以它们适合存储较长时间内需要的结果，以及保存将运行多次的程序的输入值。读者以前应该用过文字处理程序文档、电子表格、图像和视频等文件。Python 程序也存储在文件中。

文件通常分为文本文件或二进制文件。文本文件仅包含使用 ASCII 或 UTF-8 等编码系统表示字符的位序列。可以使用任何文本编辑器查看和修改这些文件。我们创建的所有 Python 程序都保存为文本文件。

与文本文件一样，二进制文件也包含位序列。但与文本文件不同的是，这些位序列可以表示任何类型的数据。它们并不仅局限于字符。包含图像、声音和视频数据的文件通常是二进制文件。本书只考虑文本文件，因为很容易用你自己最喜欢的编辑器创建和查看它们。为文本文件描述的大多数原则也可以应用于二进制文件。

7.1 打开文件

从文件中读取数据值之前，必须先打开文件。将新的数据值写入文件之前，打开文件也是必要的。通过调用 open() 函数来打开文件。

open() 函数有两个参数。第一个参数是一个字符串，包含要打开的文件的名称。第二个参数也是一个字符串，指示文件的访问模式。这里讨论的访问模式包括 read(用 r 表示)、write(用 w 表示)和 append(用 a 表示)。

文件对象由 open() 函数返回。因此，open() 函数通常在赋值语句的右侧调用，如下所示：

```
inf = open("input.txt", "r")
```

打开文件后，可将方法应用于 file 对象，以便从文件中读取数据。类似地，通过对 file 对象应用适当方法，将数据写入文件。这些方法在下一节中描述。一旦读取或写入所有的值，应该关闭文件。这是通过对 file 对象应用 close 方法实现的。

7.2 从文件中读取输入

有几种方法可以应用于 file 对象，以从文件中读取数据。这些

第 7 章 文件和异常

方法只能在以读模式打开文件时应用。试图在写入模式或追加模式下打开的文件中读取，将导致程序崩溃。

readline()方法从文件中读取一行，并将其作为字符串返回，这与 input()函数读取键盘上输入的一行文本非常相似。对 readline()的每个后续调用都从文件的顶部到底部依次读取另一行。当没有其他数据要从文件中读取时，readline()方法返回一个空字符串。

考虑一个包含一长串数字的数据文件，每个数字都独占一行。下面的程序计算该文件中所有数字的总和。

```
# Read the file name from the user and open the file
fname = input("Enter the file name: ")
inf = open(fname, "r")

# Initialize the total
total = 0

# Total the values in the file
line = inf.readline()
while line != "":
  total = total + float(line)
  line = inf.readline()

# Close the file
inf.close()

# Display the result
print("The total of the values in", fname, "is", total)
```

这个程序首先从用户处读取文件的名称。读取名称后，打开文件进行读取，file 对象存储在 inf 中，然后 total 初始化为 0，从文件中读取第一行。

然后计算 while 循环上的条件。如果从文件中读取的第一行非空，则执行循环体。它将文件中读取的行转换为浮点数，并将其加到 total 中。然后从文件中读取下一行。如果文件包含更多数据，那么 line 变量将包含文件中的下一行，while 循环条件的值计算为 True，再次执行循环，将另一个值添加到 total 中。

在某个时候，所有数据都从文件中读取出来。发生这种情况时，

readline()方法返回一个空字符串,该字符串存储到 line 中。这将导致 while 循环上的条件求值为 False,于是循环终止。然后程序继续,显示总数。

有时一次从一个文件中读取所有数据,而不是一行一行地读取数据是有帮助的。这可以使用 read()方法或 readlines()方法来完成。read()方法以一个(可能非常长的)字符串的形式返回文件的全部内容。然后进行进一步的处理,将字符串分成更小的块。readlines()方法返回一个列表,其中每个元素都是文件中的一行。用 readlines()读取所有行后,可使用循环来处理列表中的每个元素。下面的程序使用 readlines()来计算一个文件中所有数字的和。它一次从文件中读取所有数据,而不是在读取时将每个数字添加到总数中。

```
# Read the file name from the user and open the file
fname = input("Enter the file name: ")
inf = open(fname, "r")

# Initialize total and read all of the lines from the file
total = 0
lines = inf.readlines()

# Total the values in the file
for line in lines:
    total = total + float(line)

# Close the file
inf.close()

# Display the result
print("The total of the values in", fname, "is", total)
```

7.3 行结束符

下面的示例使用 readline()方法读取和显示文件中的所有行。打印时,每行前面都有行号和冒号。

```
# Read the file name from the user and open the file
fname = input("Enter the name of a file to display: ")
```

第 7 章 文件和异常

```
inf = open(fname, "r")

# Initialize the line number
num = 1

# Display each line in the file, preceded by its line number
line = inf.readline()
while line != "":
  print("%d: %s" % (i, line))

  # Increment the line number and read the next line
  num = num + 1
  line = inf.readline()

# Close the file
inf.close()
```

运行这个程序时,你可能对它的输出感到惊讶。特别是,每次打印文件中的一行时,紧接着打印的第二行是空白的。这是因为文本文件中的每一行都以一个或多个表示行末的字符结束[1]。需要这样的字符,以便读取文件的任何程序都可确定一行的结束和下一行的开始。如果没有它们,当程序读取文本文件中的所有字符(或将它们加载到文本编辑器中)时,它们将出现在同一行中。

通过调用 rstrip()方法,可从文件读取的字符串中删除行尾标记。这个方法可以应用于任何字符串,它从字符串的右端删除任何空白字符(空格、制表符和行结束符)。该方法返回删除了这些字符(如果有)的字符串的新副本。

行号程序的更新版本如下所示。它使用 rstrip()方法来删除行末标记,因此不包括前一个版本中显示不正确的空白行。

```
# Read the file name from the user and open the file
fname = input("Enter the name of a file to display: ")
inf = open(fname, "r")

# Initialize the line number
num = 1
```

---

[1] 在不同操作系统中,用来表示文本文件中一行结束的字符或字符序列是不同的。幸运的是,Python 自动处理这些差异,并允许在一种操作系统上创建的文本文件由运行在另一种操作系统上的 Python 程序加载。

```
# Display each line in the file, preceded by its line number
line = inf.readline()
while line != "":
    # Remove the end of line marker and display the line preceded by its line number
    line = line.rstrip()
    print("%d: %s" % (i, line))

    # Increment the line number and read the next line
    num = num + 1
    line = inf.readline()

# Close the file
inf.close()
```

7.4 将输出写入文件

以写入模式打开文件时,将创建一个新的空文件。如果文件已经存在,那么现有文件将被销毁,其中包含的任何数据都将丢失。如果以追加模式打开已有的文件,则写入文件的任何数据都会添加到文件末尾。如果以附加模式打开的文件不存在,则创建一个新的空文件。

write()方法可用于将数据写入以写入模式或追加模式打开的文件。它有一个必须是字符串的参数,即要写入文件的字符串。其他类型的值可调用 str()函数转换为字符串。要将多个值写入文件,可将多个项连接成一个更长的字符串,或多次调用 write()方法。

与 print()函数不同,write()方法不会在写入值后,自动移动到下一行。因此,必须在文件中将行末标记显式地写入不同行上的值之间。Python 使用\n 表示行末标记。这对字符称为转义序列,可以单独出现在字符串中,也可作为更长字符串的一部分。

下面的程序将 1 到用户输入的数字(含)写入文件。使用字符串连接和\n 转义序列,以便每个数字都写在单独一行上。

```
# Read the file name from the user and open the file
fname = input("Where will the numbers will be stored? ")
outf = open(fname, "w")
```

```
# Read the maximum value that will be written
limit = int(input("What is the maximum value? "))

# Write the numbers to the file with one number on each line
for num in range(1, limit + 1):
  outf.write(str(num) + "\n")

# Close the file
outf.close()
```

7.5 命令行参数

计算机程序通常是通过单击图标或从菜单中选择一项来执行的。也可以通过在终端或命令提示符窗口中输入适当的命令，从命令行启动程序。例如在许多操作系统中，可以通过在这样的窗口中键入 test.py 或 Python test.py，来执行存储在 test.py 中的 Python 程序。

从命令行启动程序提供了向其提供输入的新机会。程序执行其任务所需的值可以是用于启动程序的命令的一部分，方法是在命令行中，在 .py 文件的名称后面包含这些值。能将输入提供为用于启动程序的命令的一部分，这在编写使用多个程序来自动化某些任务的脚本以及编写计划定期运行的程序时特别有用。

程序执行时提供的任何命令行参数都存储在 sys(system)模块的 argv(参数向量)变量中。这个变量是一个列表，列表中的每个元素都是一个字符串。通过调用适当的类型转换函数(如 int()和 float())，可以将列表中的元素转换为其他类型。参数向量中的第一个元素是正在执行的 Python 源文件的名称。列表中的后续元素是命令行中 Python 文件名称后面提供的值(如果有)。

下面的程序演示如何访问参数向量。它首先报告提供给程序的命令行参数数量和正在执行的源文件的名称。然后，如果提供了这些值，它将继续执行，显示出现在源文件名称之后的参数。否则，将显示一条消息，表明在执行的 .py 文件之后没有命令行参数。

```
# The system module must be imported to access the command line arguments
import sys

# Display the number of command line arguments (including the .py file)
print("The program has", len(sys.argv), \
      "command line argument(s).")

# Display the name of the .py file
print("The name of the .py file is", sys.argv[0])

# Determine whether or not there are additional arguments to display
if len(sys.argv) > 1:
  # Display all of the command line arguments beyond the name of the .py file
  print("The remaining arguments are:")
  for i in range(1, len(sys.argv)):
    print(" ", sys.argv[i])
else:
  print("No additional arguments were provided.")
```

命令行参数可用于向程序提供可在命令行上输入的任何输入值，如整数、浮点数和字符串。然后，可以像使用程序中的任何其他值一样使用这些值。例如，以下代码行是前述程序的修订版本，它将文件中的所有数字相加。在这个版本的程序中，文件名作为命令行参数提供，而不是从键盘读取。

```
# Import the system module
import sys

# Ensure that the program was started with one command line argument beyond the name
# of the .py file
if len(sys.argv) != 2:
  print("A file name must be provided as a command line", \
        "argument.")
  quit()

# Open the file listed immediately after the .py file on the command line
inf = open(sys.argv[1], "r")

# Initialize the total
total = 0

# Total the values in the file
line = inf.readline()
while line != "":
  total = total + float(line)
```

```
        line = inf.readline()

# Close the file
inf.close()

# Display the result
print("The total of the values in", sys.argv[1], "is", total)
```

7.6 异常

运行程序时,可能有许多错误:期望提供数字值时,用户可能提供了一个非数字值;用户可能输入一个值,导致程序出现除 0 错误;或者用户尝试打开一个不存在的文件,等等。所有这些错误都是异常。默认情况下,在发生异常时,Python 程序会崩溃。但可通过捕获异常,并采取适当的恢复操作,来防止程序崩溃。

为了捕获异常,程序员必须指出异常可能发生的位置。还必须指出在异常发生时要运行什么代码来处理异常。这些任务是通过使用两个尚未介绍的关键字完成的:try 和 except。可能导致想要捕获的异常的代码放在一个 try 块中。try 块后面紧跟着一个或多个 except 块。当一个 try 块中发生异常时,执行流会立即跳转到相应的 except 块,而不会运行 try 块中的任何剩余语句。

每个 except 块都可指定它捕获的特定异常。为此,在 except 关键字之后紧接着异常的类型。这样的块只在指定类型的异常发生时执行。没有指定特定异常的 except 块将捕获任何类型的异常(与同一 try 块关联的另一个 except 块不会捕获它)。except 块只在发生异常时执行。如果 try 块执行时没有引发异常,那么跳过所有的 except 块,执行流继续执行最后一个 except 块之后的第一行代码。

当用户提供了一个不存在的文件名时,前几节中创建的程序都会崩溃。发生此崩溃是因为引发了未捕获的 FileNotFoundError 异常。下面的代码段使用 try 块和 except 块来捕获这个异常,并在发生该异常时显示有意义的错误消息。这个代码段后面可以跟着读取和处理文件中的数据所需的任何附加代码。

```
# Read the file name from the user
fname = input("Enter the file name: ")

# Attempt to open the file
try:
  inf = open(fname, "r")
except FileNotFoundError:
  # Display an error message and quit if the file was not opened successfully
  print("'%s' could not be opened. Quitting...")
  quit()
```

当用户请求的文件不存在时，程序的当前版本退出。虽然在某些情况下这可能没问题，但在其他情况下，提示用户重新输入文件名是更好的选择。用户输入的第二个文件名也可能导致异常。因此，必须使用循环，直到用户输入成功打开的文件名为止。下面的程序演示了这一点。注意，try 块和 except 块都在 while 循环中。

```
# Read the file name from the user
fname = input("Enter the file name: ")
file_opened = False
while file_opened == False:
  # Attempt to open the file
  try:
    inf = open(fname, "r")
    file_opened = True
  except FileNotFoundError:
    # Display an error message and read another file name if the file was not
    # opened successfully
    print("'%s' wasn't found. Please try again.")
    fname = input("Enter the file name: ")
```

当这个程序运行时，它首先从用户处读取一个文件的名称。然后 file_open 变量设置为 False，循环第一次运行。两行代码驻留在循环体的 try 块中。第一次尝试打开用户指定的文件。如果文件不存在，则引发 FileNotFoundError 异常，执行流立即跳到 except 块，跳过 try 块中的第二行。当 except 块执行时，它会显示一条错误消息，并从用户处读取另一个文件名。

执行流继续返回到循环的顶部，再次评估它的条件。该条件仍然计算为 False，因为 file_open 变量仍然为 False。结果，循环体第二次执行，程序再次尝试使用最近输入的文件名打开文件。如果该文件不存在，则程序将按照前一段所述进行。但是，如果文件存在，对 open 的调用将成功完成，执行流将继续执行 try 块中的下一行。这一行将 file_open 设置为 True。然后跳过 except 块，因为在执行 try 块时没有引发任何异常。最后，循环终止，因为 file_open 设置为 True，程序的其余部分继续执行。

本节介绍的概念可用于检测和响应程序运行过程中可能出现的各种错误。通过创建 try 和 except 块，程序可用适当方式响应这些错误，而不会崩溃。

7.7 练习

本章的许多练习都是从文件中读取数据。在某些情况下，任何文本文件都可以用作输入。在其他情况下，可以在自己喜欢的文本编辑器中轻松地创建适当的输入文件。还有一些练习需要特定的数据集，如单词、名称或化学元素的列表。这些数据集可从作者的网站下载：

http://www.cpsc.ucalgary.ca/~bdstephe/PythonWorkbook

练习 149：显示文件的头部

(有解答，40 行)

基于 UNIX 的操作系统通常包括一个名为 head 的工具。把文件名提供为命令行参数，该工具显示文件的前 10 行。编写一个提供相同行为的 Python 程序。如果用户请求的文件不存在，或者省略了命令行参数，则显示适当的错误消息。

练习 150：显示文件的尾部

(有解答，35 行)

基于 UNIX 的操作系统通常还包括一个名为 tail 的工具。把文件名提供为命令行参数，该工具显示文件的后 10 行。编写一个提供相同行为的 Python 程序。如果用户请求的文件不存在，或者省略了命令行参数，则显示适当的错误消息。

有几种不同的方法可用来解决这个问题。一种选择是将文件的全部内容加载到列表中，然后显示最后 10 个元素。另一个选项是读取文件的内容两次，一次用来计算行数，另一次用来显示最后 10 行。然而，这两种解决方案在处理大型文件时都不合适。另一种解决方案只需要读取文件一次，并且一次只需要存储文件中的 10 行。开发这样一个解决方案，来完成这个额外挑战。

练习 151：连接多个文件

(有解答，28 行)

基于 UNIX 的操作系统通常包括一个名为 cat 的工具，它是 concatenate 的缩写。把一个或多个文件名作为命令行参数提供给该工具，该工具就会显示这些文件的内容连接起来的结果。文件的显示顺序与它们在命令行中显示的顺序相同。

创建执行此任务的 Python 程序。它应该为任何不能显示的文件生成适当的错误消息，然后继续处理下一个文件。如果程序启动时没有任何命令行参数，则显示适当的错误消息。

练习 152：给文件中的行编号

(23 行)

创建一个程序，从文件中读取行，向其中添加行号，然后将编号的行存储到新文件中。输入文件的名称从用户处读取，程序创建的新文件的名称也是如此。输出文件中的每一行都应该以行号开头，后跟冒号和空格，然后是输入文件中的行。

练习 153：找出文件中最长的单词

(39 行)

本练习将创建一个 Python 程序来识别文件中最长的单词。程序应该输出适当的消息，其中包括最长单词的长度，以及文件中出现的该长度的所有单词。将任何一组非空白字符视为一个单词，即使它包含数字或标点符号。

练习 154：字母频率

(43 行)

频率分析是一种可用来帮助破解一些简单加密形式的技术。这个分析检查加密的文本，以确定哪些字符是最常见的。然后尝试将英语中最常见的字母(如 E 和 T)映射到加密文本中最常见的字符。

编写一个程序，来确定和显示文件中所有字母的频率。在执行此分析时，请忽略空格、标点符号和数字。程序应该不区分大小写，将 a 和 A 视为等同。用户提供要分析的文件的名称作为命令行参数。如果用户提供了错误数量的命令行参数，或者如果程序无法打开用户指定的文件，则程序应该显示有意义的错误消息。

练习 155：最常出现的单词

(37 行)

编写一个程序，来显示文件中出现频率最高的单词。程序应该首先读取用户输入的文件名。然后应该处理文件中的每一行。每一行都需要分成单词，并且需要从每个单词中删除任何开头或结尾的标点符号。在计算每个单词出现的次数时，程序也应该忽略大小写。

> 提示：完成这个任务时，会发现练习 117 的答案很有帮助。

练习 156：对一组数字求和

(有解答，26 行)

创建一个程序，来汇总用户输入的所有数字，同时忽略任何不是有效数字的输入。程序应该在输入每个数字后显示当前的总和。它应该在每个非数字输入之后显示适当的消息，然后继续对用户输入的任何其他数字求和。当用户输入空行时退出程序。确保程序对整数和浮点数都能正确工作。

> **提示**：这个练习要求在不使用文件的情况下使用异常。

练习 157：字母成绩和绩点

(106 行)

编写一个程序，从字母等级和绩点之间来回转换。程序应该允许用户转换多个值，每行输入一个值。首先尝试将用户输入的每个值从绩点转换为字母等级。如果在此过程中出现异常，则程序应尝试将该值从字母级别转换为若干个绩点。如果两个转换都失败，则程序应该输出一条消息，指示提供的输入无效。设计程序，使它继续执行转换(或报告错误)，直到用户输入空行为止。完成这个练习时，练习 52 和 53 的解答可能会有帮助。

练习 158：删除注释

(有解答，53 行)

Python 使用#字符来标记注释的开头。注释从#字符一直延续到包含它的行尾。Python 没有提供任何在行末之前结束注释的机制。

本练习将创建一个从 Python 源文件中删除所有注释的程序。检查文件中的每一行以确定是否存在#字符。如果是，那么程序应该删除从#字符到行尾的所有字符(忽略注释字符出现在字符串内部的情况)。使用新名称保存修改后的文件。输入文件的名称和输出文件的名称都应该从用户处读取。确保在访问任何一个文件时，若遇到问题，则显示适当的错误消息。

练习159：两个单词的随机密码

(有解答，39 行)

虽然通过选择随机字符来生成密码通常会创建一个相对安全的密码，但密码通常很难记住。一种替代方法是，一些系统通过获取两个英语单词并将它们连接起来，来构造密码。虽然这个密码可能不那么安全，但通常更容易记住。

编写一个程序，它读取一个包含单词列表的文件，随机选择其中的两个单词，并将它们连接起来生成一个新密码。在生成密码时，请确保总长度为8～10个字符，并且使用的每个单词至少有三个字母长。将密码中的每个单词大写，这样用户就可以很容易看出一个单词的结尾和下一个单词的开头。最后，程序应该给用户显示密码。

练习160：奇怪的单词

(67 行)

学英语拼写的学生经常被教押韵"i 在 e 之前，c 之后则反向"。这个经验法则建议，当单词中的 i 和 e 相邻时，i 应在 e 的前面；除非它们紧跟在 c 后面，当 c 在 e 的前面时，e 就在 i 的前面。这些建议适用于 e、i 的前面没有 c 的单词，如 believe、chief、fierce 和 friend；同样适用于 e、i 紧跟在 c 后面的单词，如 ceiling 和 receipt。但这个规则也有例外，如 weird。

创建一个程序，来处理包含文本行的文件。文件中的每一行可能包含许多单词(或者根本没有单词)。应该忽略任何不包含 e 与 i 相邻的单词。应该检查包含 e 和 i 相邻的单词(按任意顺序)，以确定它们是否遵循"i 在 e 之前，c 之后则反向"规则。构造并报告两个列表：一个包含所有遵循该规则的单词，另一个包含所有违反该规则的单词。两个列表都不应该包含任何重复的值。在程序结束时报告列表的长度，这样就可以很容易地确定文件中符合"i 在 e

之前，c 之后则反向"规则的单词比例。

练习 161：那个元素是什么？

(59 行)

编写一个程序，读取包含化学元素信息的文件，并将其存储在一个或多个适当的数据结构中。然后，程序应该读取并处理来自用户的输入。如果用户输入一个整数，那么程序应该显示输入质子数的元素的符号和名称。如果用户输入一个非整数值，那么程序应该为该名称或符号的元素显示质子数。如果输入的名称、符号或质子数不存在对应的元素，则程序应显示适当的错误消息。继续从用户处读取输入，直到输入空行为止。

练习 162：一本没有 E 的书……

(有解答，50 行)

小说《葛士比》有 5 万多字长。对于一部小说来说，5 万个单词并不奇怪，但该书中没有一个单词使用字母 E。考虑到 E 是英语中最常见的字母，这一点尤其值得注意。

编写一个程序，从一个文件中读取单词列表，并确定单词使用字母表中每个字母的比例。给所有 26 个字母显示结果，并包含一个额外消息，来标识单词中使用比例最小的字母。程序应该忽略文件中出现的任何标点符号，并将大写字母和小写字母视为等同。

> 避讳某字母之文是一种不使用特定字母(或一组字母)的书面作品。要避免的字母通常是元音字母，但有时是辅音字母。例如，埃德加·爱伦·坡的《乌鸦》是一首超过 1000 个单词的诗，它不使用字母 Z，因此是一个避讳某字母之文。*La Disparition* 是避讳某字母之文的另一个例子。原小说(用法语写的)和它的英文译本 *A Void* 大约 300 页，除了作者的名字之外，没有使用字母 E。

练习 163：排名第一的名字

(有解答，54 行)

婴儿姓名数据集由 200 多个文件组成。每个文件都有一个包含 100 个姓名的列表，以及使用每个姓名的次数。文件中的条目是按从最常用到最不常用的顺序排列的。每年有两个文件：一个包含女孩的姓名，另一个包含男孩的姓名。数据集包括 1900 年到 2012 年的数据。

编写一个程序，读取数据集中的每个文件，并识别至少一年内最流行的所有姓名。程序应该输出两个列表：一个包含最受欢迎的男孩姓名，另一个包含最受欢迎的女孩姓名。两个列表都不应该包含任何重复的值。

练习 164：中性名字

(56 行)

一些名字，如 Ben、Jonathan 和 Andrew 通常只用于男孩，而 Rebecca 和 Flora 等名字通常只用于女孩。其他名字，如 Chris 和 Alex，可能男女通用。

编写一个程序，确定并显示在用户指定的年份中所有男孩和女孩使用的婴儿名。如果在选定的年份中没有中性名字，程序应该生成适当的消息。如果用户请求的年份没有数据，则显示适当的错误消息。关于婴儿姓名数据集的更多细节包含在练习 163 中。

练习 165：特定时期内出生的婴儿的最常用姓名

(76 行)

编写一个程序，使用练习 163 中描述的婴儿姓名数据集，来确定在一段时间内哪些姓名用得最频繁。用户提供范围的始末年份进行分析。显示指定时期使用最多的男孩姓名和女孩姓名。

练习 166：不同的姓名

(41 行)

这个练习将创建一个程序，来读取练习 163 中描述的婴儿姓名数据集中的每个文件。当程序读取文件时，它应该跟踪男孩和女孩使用的每个不同姓名。然后，程序应该输出这些姓名的列表。两个列表都不应该包含任何重复的值。

练习 167：拼写检查

(有解答，58 行)

拼写检查器对于拼写错误的人来说是一个很有用的工具。本练习将编写一个程序，来读取文件，并显示其中所有拼错的单词。根据已知单词列表检查文件中的每个单词，来识别拼写错误的单词。用户文件中没有出现在已知单词列表中的任何单词都将报告为拼写错误。

用户将提供文件名作为命令行参数来检查拼写错误。如果缺少命令行参数，程序应该显示适当的错误消息。如果程序无法打开用户的文件，还应该显示一条错误消息。解答这个练习时，可以使用练习 117 的答案，这样逗号、句号或其他标点符号后面的单词就不会报告为拼写错误。检查拼写时忽略单词的大小写。

> **提示**：虽然可将单词数据集中的所有英语单词加载到一个列表中，但是如果使用 Python 的 in 操作符，则搜索列表的速度会很慢。检查某键是否存在于字典中，或者某值是否存在于集合中要快得多。如果使用字典，单词就是键。这些值可以是整数 0(或任何其他值)，因为这些值永远不会使用。

练习 168：重复的单词

(61 行)

拼写错误只是书面作品中可能出现的多种错误之一。另一个常

见错误是重复的单词。例如,作者可能无意中复制了一个单词,如下面的句子所示:

```
At least one value must be entered
entered in order to compute the average.
```

一些文字处理程序会检测出这个错误,并在执行拼写或语法检查时识别它。

本练习将编写一个程序,来检测文本文件中的重复单词。当找到重复单词时,程序应该显示包含行号和重复单词的消息。确保程序正确处理同一单词出现在一行末尾和下一行开头的情况,如前面的示例所示。要检查的文件名称将作为程序的唯一命令行参数提供。如果用户未能提供命令行参数,或者在处理文件时发生错误,则显示适当的错误消息。

练习169:编辑文件中的文本

(有解答,52行)

敏感信息在向公众发布之前,通常会从文件中删除或修改。当文档发布时,通常会用黑条替换修订后的文本。

本练习将编写一个程序,用星号替换文本文件中出现的所有敏感单词,从而对它们进行修订。你的程序应该对出现在任何地方的敏感词汇进行修订,即使它们出现在另一个单词的中间。敏感词的列表在单独的文本文件中提供。将经过修订的原始文本保存到新文件中。原始文本文件、敏感词文件和修订文件的名称都由用户提供。

> 在完成这个练习时,可能发现字符串的 replace()方法非常有用。有关 replace()方法的信息可在因特网上找到。
>
> 对于额外的挑战,扩展程序,使它不区分大小写的方式编辑单词。例如,如果 exam 出现在敏感词列表中,那么请对 exam、Exam、ExaM 和 EXAM 等可能的形式进行编辑。

练习 170：缺少注释

(有解答，48 行)

在编写函数时，最好包含一条注释，概述函数的用途、参数和返回值。然而，有时会遗忘注释，或者被一些好心的程序员遗漏了，这些程序员打算以后再写注释，但从来没有时间去写。

创建一个 Python 程序，该程序读取一个或多个 Python 源文件，并标识没有紧跟在注释之后的函数。出于本练习的目的，假设以 def 开头，后跟空格的任何行都是函数定义的开头。假设当函数有注释时，注释字符#将是前一行的第一个字符。显示所有缺少注释的函数名称，以及函数定义所在的文件名和行号。

用户提供一个或多个 Python 文件的名称作为命令行参数，所有这些参数都应该由程序进行分析。对于不存在或无法打开的任何文件，应该显示适当的错误消息。然后，程序应该处理其余文件。

练习 171：一致的行长度

(45 行)

对于终端窗口来说，80 个字符是常见的宽度，而有些终端窗口略窄或略宽。在显示包含文本段落的文档时，这可能会带来挑战。这些行可能太长，很难阅读，或者太短，不能很好地利用可用的空间。

编写一个程序，打开一个文件并显示它，使每一行都尽可能满。如果读取的一行太长，那么程序应该把它分成单词，并把它们添加到当前行中，直到写满为止。然后，程序应该开始一个新行，并显示其余的单词。类似地，如果读取的行太短，则需要使用文件下一行中的单词，来填充当前输出行。例如，考虑一个文件，其中包含"爱丽丝梦游仙境"中的以下几行：

```
Alice was
beginning to get very tired of sitting by her
sister
on the bank, and of having nothing to do: once
```

```
or twice she had peeped into the book her sister
was reading, but it had
no
pictures or conversations in it,"and what is
the use of a book," thought Alice, "without
pictures or conversations?"
```

当格式化为50个字符的行长度时，它应该显示为：

```
Alice was beginning to get very tired of sitting
by her sister on the bank, and of having nothing
to do: once or twice she had peeped into the book
her sister was reading, but it had no pictures or
conversations in it, "and what is the use of a
book," thought Alice, "without pictures or
conversations?"
```

确保程序对包含多个文本段落的文件正确工作。可以检测一个段落的结束和下一个段落的开始，方法是在移除行结束标记之后查找空行。

> **提示**：使用常量表示最大的行长度。需要不同的行长度时，将更容易更新程序。

练习172：按顺序排列六个元音字母的单词

(56行)

在英语中，哪些单词包含每个元音字母A、E、I、O、U和Y(而且是按顺序一次包含的)？编写一个程序，搜索一个包含单词列表的文件，并显示满足此约束的所有单词。用户提供要搜索的文件的名称。如果用户提供了一个无效的文件名，或者当程序按顺序搜索六个元音字母的单词时出现了其他错误，就显示适当的错误消息并退出程序。

第8章 递　　归

　　第4章探讨了函数的许多方面，包括调用其他函数的函数。之前没有考虑的一个密切相关的主题是函数是否可以调用自身。事实证明，这是可能的，而且它是解决一些问题的强大技术。

　　用自身来描述事物的定义称为递归。递归定义要想有用，就必须用不同的(通常是更小或更简单的)版本来描述所定义的内容。用相同版本的自身定义某个东西，虽然是递归的，但并不是特别有用，因为这个定义是循环的。有用的递归定义必须使用已知的解向问题的版本发展。

　　任何调用自身的函数都是递归的，因为函数的主体(它的定义)包含了对被定义函数的调用。为得到解，递归函数必须至少在一种情况下能产生所需的结果，而不需要调用自身。这称为基本用例。函数调用自身的情况称为递归情况。下面用三个例子讨论递归。

8.1 整数求和

考虑计算从 0 到 n 的所有整数的和的问题。这可以使用循环或公式来完成。它也可以递归地执行。最简单的情况是 n = 0。这种情况下，已知答案是 0，可以返回该答案，而不需要使用问题的另一个版本。因此，这是基本用例。

对于任何正整数 n，要计算从 0 到 n(含)的和，可将 n 加到从 0 到 n‐1(含)的和上。这个描述是递归的，因为 n 个整数的和表示为同一个问题的更小版本(从 0 到 n‐1 求和，并包含 n‐1)，加上少量的额外工作(将 n 添加到该和)。每次应用这个递归定义时，它都向基本用例更进一步(n 为 0 时)。当达到基本用例时，不再执行递归。这允许计算完成并返回结果。

下面的程序实现了前面段落中描述的递归算法，计算从 0 到用户输入的正整数(含)的整数和。if 语句用于确定是执行基本用例还是递归用例。当基本用例执行时，立即返回 0，而不进行递归调用。当递归情况执行时，会用一个较小参数(n‐1)再次调用函数。一旦递归调用返回，n 会添加到返回的值中。这成功地计算从 0 到 n(含)的所有值，然后将这些值作为函数的结果返回。

```
# Compute the sum of the integers from 0 up to and including n using recursion
# @param n the maximum value to include in the sum
# @return the sum of the integers from 0 up to and including n
def sum_to(n):
    if n <= 0:
        return 0                       # Base case
    else:
        return n + sum_to(n - 1) # Recursive case

# Compute the sum of the integers from 0 up to and including a value entered by the user
num = int(input("Enter a non-negative integer: "))
total = sum_to(num)
print("The total of the integers from 0 up to and including", \
      num, "is", total)
```

考虑如果用户在程序运行时输入 2 会发生什么。此值从键盘读取,并转换为整数。然后以 2 作为参数调用 sum_to()。if 语句的条件计算为 True,因此执行它的主体,递归调用 sum_to(),其参数为 1。这个递归调用必须在计算 sum_to() 的副本(其中 n = 2)并返回结果之前完成。

执行对 sum_to() 的递归调用(其中 n = 1)将导致对 sum_to() 的另一个递归调用,其中 n = 0。一旦这个递归调用开始执行,就会执行带有参数值 2、1 和 0 的 sum_to() 的三个副本。n=2 的 sum_to() 副本等待 n=1 的 sum_to() 副本完成,才可以返回结果,n=1 的 sum_to() 副本等待 n=0 的 sum_to() 副本完成,才可以返回结果。虽然所有函数都有相同的名称,但是函数的每个副本都完全独立于所有其他副本。

当 n = 0 的 sum_to() 副本执行时,执行基本用例。它立即返回 0。将递归调用返回的 0 加上 1,就允许继续执行 n = 1 的 sum_to() 副本。然后调用 n = 1 时返回的总数 1。继续执行 n = 2 的 sum_to() 副本。它将 n(即 2)添加到递归调用返回的 1 上,返回 3,并存储到 total 中。最后,显示总数,程序终止。

8.2 斐波那契数

斐波那契数列是一个从 0 和 1 开始的整数序列。序列中随后的每个数都是其前两个数的和。因此,斐波那契数列的前 10 个数字是 0、1、1、2、3、5、8、13、21 和 34。斐波那契数列中的数字通常用 F_n 表示,其中 n 是一个非负整数,用于标识数列中的数字索引(从 0 开始)。

斐波那契数列中的数字,除了前两个以外,可以用公式 $F_n = F_{n-1}+F_{n-2}$ 来计算。这个定义是递归的,因为较大的斐波那契数是用两个较小的斐波那契数来计算的。序列中的前两个数字 F_0 和 F_1 是基本用例,因为它们具有未进行递归计算的已知值。下面的程序实现了计算斐波那契数列的递归公式。对于用户输入的某个 n 值,它

计算并显示 F_n 值。

```
# Compute the nth Fibonacci number using recursion
# @param n the index of the Fibonacci number to compute
# @return the nth Fibonacci number
def fib(n):
  # Base cases
  if n == 0:
    return 0
  if n == 1:
    return 1

  # Recursive case
  return fib(n-1) + fib(n-2)

# Compute the Fibonacci number requested by the user
n = int(input("Enter a non-negative integer: "))
print("fib(%d) is %d." % (n, fib(n)))
```

这种用于计算斐波那契数列的递归算法比较紧凑，但速度很慢，即使处理的是相当小的值。虽然现代计算机将很快返回计算 fib(35) 的结果，但计算 fib(70) 将需要几年时间。因此，较大的斐波那契数通常使用循环或公式计算。

基于斐波那契数列程序的性能，可能得出这样的结论：递归解决方案太慢了，无法发挥作用。虽然在这种特殊情况下是这样，但一般情况下就不一样了。在之前的项目中，整数和的运算速度很快，甚至对于更大的值也是如此，而且有些问题有非常有效的递归算法，例如欧几里得算法，练习 174 使用该算法计算两个整数的最大公约数。

下图演示了计算 F_4 和 F_5 时的递归调用，以及计算 sum_to(4) 和 sum_to(5) 的递归调用。比较为计算不同输入值的结果而进行的函数调用，可以看出这些问题在效率上的差异。

当传递给 sum_to() 的参数从 4 增加到 5 时，函数调用的数量也从 4 增加到 5。更一般地，当传递给 sum_to() 的参数增加 1 时，函数调用的数量也增加 1。这称为线性增长，因为递归调用的数量与第一次调用函数时提供的参数值成正比。

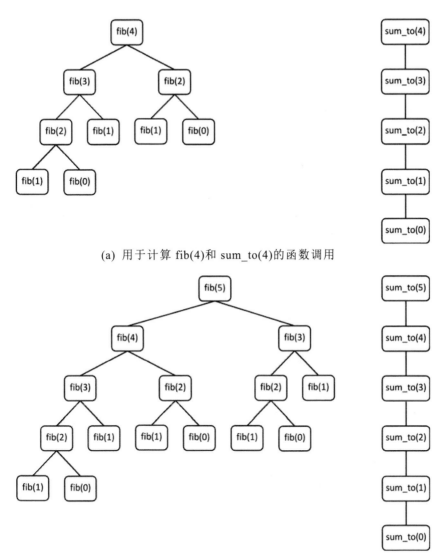

(a) 用于计算 fib(4)和 sum_to(4)的函数调用

(b) 用于计算 fib(5)和 sum_to(5)的函数调用

相反,当传递给 fib()的参数从 4 增加到 5 时,函数调用的数量从 9 增加到 15。更一般地说,当计算的斐波那契数列的位置增加 1 时,递归调用的数量几乎增加一倍。这称为指数增长。指数增长使得在任何实际意义上计算大值的结果变得不可能,因为重复地将计算所需的时间翻番,会导致运行时间太长而无实际用途。

8.3 计算字符个数

递归可以用来解决任何可用自身表示的问题。它不局限于整数操作问题。例如,考虑如下问题:计算特定字符 ch 在字符串 s 中出现的次数。可以编写一个递归函数来解决这个问题,该函数将 s 和 ch 作为参数,返回 ch 在 s 中出现的次数作为其结果。

这个问题的基本用例是:s 是空字符串。由于空字符串不包含任何字符,所以它包含 ch 的次数为 0。因此,在这种情况下,函数可以返回 0,而不需要进行递归调用。

ch 在较长的字符串中出现的次数可通过以下递归方式确定。为简化递归情况的描述,将 s 的尾部定义为 s 中除第一个字符外的所有字符。只包含一个字符的字符串尾部是空字符串。

如果 s 中的第一个字符是 ch,则 s 中 ch 出现的次数是 1 加 ch 在 s 尾部出现的次数,否则 ch 在 s 中出现的次数就是 ch 在 s 尾部出现的次数。这个定义使处理过程趋向基本用例(s 是空字符串),因为 s 尾部总是短于 s。实现这个递归算法的程序如下所示。

```
# Count the number of times a particular character is present in a string
# @param s the string in which the characters are counted
# @param ch the character to count
# @return the number of occurrences of ch in s
def count(s, ch):
  if s == "":
    return 0 # Base case

  # Compute the tail of s
  tail = s[1 : len(s)]

  # Recursive cases
  if ch == s[0]:
    return 1 + count(tail, ch)
```

```
else:
    return count(tail, ch)
```

```
# Count the number of times a character entered by the user occurs in a string entered
# by the user
s = input("Enter a string: ")
ch = input("Enter the character to count: ")
print("'%s' occurs %d times in '%s'" % (ch, count(s, ch), s))
```

8.4 练习

递归函数是调用自身的函数。这种函数通常包括一个或多个基本用例和一个或多个递归用例。当基本用例执行时，不必进行递归调用即可计算函数的结果。递归用例通过一次或多次递归调用来计算结果，通常是对问题的更小或更简单的版本进行调用。本章中的所有练习都应该通过编写一个或多个递归函数来解决。这些函数中的每一个都将调用自己，还可利用前几章中讨论的 Python 特性。

练习 173：把这些值加起来

(有解答，29 行)

编写一个程序，从用户处读取值，直到输入空行为止。显示用户输入的所有值的总和(如果输入的第一个值是空行，则为 0.0)。使用递归完成此任务。程序可能不使用任何循环。

> 提示：递归函数的主体需要从用户处读取一个值，然后决定是否进行递归调用。函数不需要接收任何参数，但需要返回一个数字结果。

练习 174：最大公约数

(24 行)

欧几里得是一位生活在大约 2300 年前的希腊数学家。他计算

两个正整数 a 和 b 的最大公约数的算法是有效的,也是递归的。现概述如下:

  if b 是 0,then
   return a
  else
   设 c 等于 a 除以 b 的余数
   return b 和 c 的最大公约数

编写一个实现欧几里得算法的程序,并使用它来确定用户输入的两个整数的最大公约数。用一些非常大的整数测试程序。即使是由数百位数字组成的庞大数字,也会很快计算出结果,因为欧几里得算法非常高效。

练习 175:十进制到二进制的递归

(34 行)

在练习 82 中编写了一个程序,该程序使用循环将十进制数转换为二进制数。本练习将使用递归执行相同的任务。

编写一个递归函数,将一个非负的十进制数转换成二进制数。将 0 和 1 视为返回包含适当数字的字符串的基本用例。对于所有其他正整数 n,应该使用余数运算符计算下一位数字,然后进行递归调用来计算 n // 2 的数字。最后,应该将递归调用的结果(是一个字符串)和下一个数字(需要将其转换为字符串)连接起来,并将此字符串作为函数的结果返回。

编写一个 main 程序,使用递归函数,将用户输入的非负整数从十进制转换为二进制。如果用户输入负值,程序应该显示适当的错误消息。

练习 176:北约语音字母

(33 行)

拼写字母表是一组单词，每个单词代表字母表中的 26 个字母中的一个。虽然在低质量或嘈杂的通信信道中，许多字母很容易被误解，但在拼写字母表中，用来表示字母的单词通常都是经过选择的，因此每个字母听起来都是不同的，很难与其他字母混淆。北约语音字母表是一种广泛使用的拼写字母表。表 8.1 显示了每个字母及其相关单词。

表 8.1　北约语音字母表

字母	单词	字母	单词	字母	单词
A	Alpha	J	Juliet	S	Sierra
B	Bravo	K	Kilo	T	Tango
C	Charlie	L	Lima	U	Uniform
D	Delta	M	Mike	V	Victor
E	Echo	N	November	W	Whiskey
F	Foxtrot	O	Oscar	X	Xray
G	Golf	P	Papa	Y	Yankee
H	Hotel	Q	Quebec	Z	Zulu
I	India	R	Romeo		

编写一个程序，从用户那里读取一个单词，然后显示它的语音拼写。例如，如果用户输入 Hello，那么程序应该输出 Hotel Echo Lima Lima Oscar。程序应该使用递归函数来执行此任务。在解的任何地方都不要使用循环。用户输入的任何非字母字符都应该忽略。

练习 177：罗马数字

(25 行)

顾名思义，罗马数字是在古罗马发展起来的。即使在罗马帝国灭亡之后，其数字在欧洲仍广泛使用，直到中世纪后期，其数字在今天仍用于有限的情况。

罗马数字由字母 M、D、C、L、X、V 和 I 组成，分别代表 1000、500、100、50、10、5 和 1。书写数字时，通常是从最大值到最小值。这种情况发生时，数字的值就是所有数字的值的和。如果较小的值在较大的值之前，那么较小的值就会从它紧挨着的那个较大的值中减去，这个差值就会加到数字的值上[1]。

创建一个将罗马数字转换为整数的递归函数。函数应该在字符串的开头处理一个或两个字符，然后对所有未处理的字符递归地调用自己。对于基本用例，使用值为 0 的空字符串。另外，编写一个 main 程序，从用户那里读取罗马数字并显示其值。可以假设用户输入的值是有效的。程序不需要做任何错误检查。

练习 178：递归回文

(有解答，30 行)

回文的概念之前在练习 75 中介绍过。这个练习将编写一个递归函数，来确定一个字符串是不是回文。空字符串是回文，任何只包含一个字符的字符串都是回文。任何 3 字符的字符串，如果它的第一个和最后一个字符相同，就是回文。通过删除字符串的第一个和最后一个字符，对形成的字符串递归调用该函数，就可以判断字符串是否是回文。

编写一个 main 程序，从用户那里读取一个字符串，并使用递归函数来确定它是不是回文。然后，程序应该为用户显示适当的消息。

练习 179：递归平方根

(20 行)

练习 74 探索了如何使用迭代来计算一个数的平方根。在这个

---

1 只有 C、X 和 I 用减法。把 C、X 或 I 作为前缀的数字值不得超过被减去的值的 10 倍。因此，可以把 V 或 X 作为前缀，但不能把 L、C、D 或 M 作为前缀。例如，99 必须表示为 XCIX，而不是 IC。

第8章 递归

练习中,每次额外的循环迭代都会生成一个更好的平方根近似值。这个练习将使用相同的近似策略,但使用递归而不是循环。

创建一个有两个参数的平方根函数。第一个参数 n 是计算平方根的数字。第二个参数 guess 是当前对平方根的猜测。guess 参数的默认值应该是 1.0。不要为第一个参数提供默认值。

平方根函数是递归的。$guess^2-n$ 处于 10^{-12} 范围时出现基本用例。这种情况下,函数应该返回 guess,因为它非常接近 n 的平方根。否则,函数应该返回调用自身的结果,其中 n 是第一个参数,$\dfrac{guess + \dfrac{n}{guess}}{2}$ 是第二个参数。

编写一个 main 程序,通过计算几个不同值的平方根来演示平方根函数。从 main 程序中调用平方根函数时,应该只向它传递一个参数,guess 则使用默认值。

练习 180:字符串编辑距离

(有解答,43 行)

两个字符串之间的编辑距离是它们相似性的度量。编辑距离越小,字符串在将一个字符串转换为另一个字符串所需的插入、删除和替换操作的最小数量方面就越相似。

考虑字符串 kitten 和 sitting。通过以下操作可将第一个字符串转换为第二个字符串:用 s 代替 k,用 i 代替 e,在字符串的末尾插入 g。这是能把 kitten 变成 sitting 的最少操作数。因此,编辑距离为 3。

编写一个递归函数,计算两个字符串之间的编辑距离。使用以下算法:

设 s 和 t 是字符串
if s 的长度是 0,then
　return　t 的长度

else if t 的长度是 0, then
  return s 的长度
else
  设 cost 为 0
  if s 中的最后一个字符不等于 t 中的最后一个字符, then
    设 cost 为 1
  设 d1 为 s 中除最后一个字符外的所有字符与 t 中所有字符之间的编辑距离,加 1
  设 d2 等于 s 中的所有字符与 t 中除最后一个字符外的所有字符之间的编辑距离,加 1
  将 d3 设为 s 中除最后一个字符外的所有字符与 t 中除最后一个字符外的所有字符之间的编辑距离,加上 cost
  return d1、d2 和 d3 的最小值

使用递归函数编写一个程序,从用户读取两个字符串,并显示它们之间的编辑距离。

练习 181：可能的改变

(41 行)

创建一个程序,确定是否可以使用特定数量的硬币构建一个特定的总数。例如,总数是 1 美元,如果全部是 25 美分,就用 4 个硬币。然而,用 5 个硬币是不可能得到 1 美元的。然而,用 3 个 25 分硬币、2 个 10 分硬币和 1 个 5 分硬币就可以得到 1 美元。同样,5 个硬币或 8 个硬币可以组成 1.25 美元,但 4 个、6 个或 7 个硬币不能组成 1.25 美元。

程序应该同时读取用户输入的金额和硬币数量。然后,它应该显示一个明确的信息,指示输入的美元金额是否可以使用所指示的硬币数量形成。在完成这道题时,假设有 25 分硬币、10 分硬币、

5分硬币和1分硬币。解答必须使用递归,不能包含任何循环。

练习182:用元素符号拼写

(67 行)

每种化学元素都有一个标准符号,该符号有一个、两个或三个字母长。有些人喜欢玩的一个游戏是确定一个单词是否可以只使用元素符号拼写。例如,硅(silicon)可以用 Si、Li、C、O 和 N 等符号来拼写。但氢(hydrogen)不能用任何元素符号的组合来拼写。

编写一个递归函数,确定一个单词是否可以只使用元素符号拼写。函数将需要两个参数:尝试拼写的单词和可以使用的符号列表。它返回一个字符串作为结果,该字符串包含用于实现拼写的符号,如果不存在拼写,则返回一个空字符串。当函数搜索拼写时,应该忽略大小写。

创建一个程序,使用函数查找和显示所有只能用元素符号拼写的元素名称。显示字母的名称和符号的顺序。例如,输出的一行是:

```
Silver can be spelled as SiLvEr
```

程序使用元素数据集,这些数据集可从作者的网站下载。这个数据集包括所有118种化学元素的名称和符号。

练习183:元素序列

(有解答,81 行)

有些人喜欢玩一种游戏,这种游戏构造一个化学元素序列,序列中的每个元素都以前一个元素的最后一个字母开始。例如,如果一个序列以氢(Hydrogen)开始,那么下一个元素必须是一个以 N 开始的元素,比如镍(Nickel)。镍之后的元素必须以 L 开头,如锂(Lithium)。构造的元素序列不能包含任何重复项。当单独游戏时,游戏的目标是构建尽可能长的元素序列。当两个玩家一起玩的时候,目标是选择一个元素,让对手没有能添加到序列中的选项。

编写一个程序，从用户那里读取元素的名称，并使用递归函数查找以该值开头的最长元素序列。一旦查找到序列，就显示它。如果用户没有输入有效的元素名，请确保程序以合理方式进行响应。

提示：程序可能需要两分钟才能找到某些元素的最长序列。因此，可能想要使用像钼(Molybdenum)和镁(Magnesium)这样的元素作为第一个测试用例。它们都有一个只包含8个元素的最长序列，程序应该在几分之一秒内找到它。

练习 184：把一张清单压平

(有解答，33行)

Python的列表可包含其他列表。当一个列表在另一个列表中时，称为内部列表嵌套在外部列表中。嵌套在外部列表中的每个内部列表也可包含嵌套列表，这些列表可以包含任意深度的附加嵌套列表。例如，下面的列表包含嵌套在几个不同深度的元素：[1, [2,3], [4, [5, [6,7]], [[[8],9],[10]]]。

包含多层嵌套的列表在描述值之间的复杂关系时非常有用，但是对其中的值执行某些操作时，这样的列表也会比较困难，因为值是嵌套在不同级别上的。压平列表是将包含多层嵌套列表的列表转换为所有元素的嵌套级别都相同的列表。例如，将前一段的列表压平，结果是[1、2、3、4、5、6、7、8、9、10]。下面的递归算法可用来压平一个名为data的列表：

   if data 为空，then
    return 空列表
   if data 中的第一个元素是一个列表，then
    将 l1 设为将 data 中的第一个元素压平的结果
    将 l2 设为除第一个元素外，data 中的所有元素都被压平的结果
    return l1 和 l2 的连接

第8章 递 归

  if data 中的第一个元素不是列表，then
    设 l1 为只包含 data 中第一个元素的列表
    设 l2 为除第一个元素外，data 中的所有元素都被压平的结果
    return l1 和 l2 的连接

编写一个函数，来实现前面描述的递归压平算法。函数将接收一个参数，即要压平的列表，并返回一个结果，即已压平的列表。包含一个 main 程序，该 main 程序演示函数成功地压平此问题前面显示的列表和其他一些列表。

> **提示**：Python 包含一个名为 type() 的函数，它返回其唯一参数的类型。有关使用此函数确定变量是否为列表的信息可以在网上找到。

练习 185：运行长度解码

(33 行)

运行长度编码是一种简单的数据压缩技术，当重复值出现在列表的相邻位置时，可有效地进行压缩。为了实现压缩，应将一组重复的值替换为该值的一个副本，后跟值的重复次数。例如，列表 ("A","A","A","A","A","A","A","A","A","A","A","A"、"B","B","B","B", "A","A","A","A","A","A","B") 压缩为 ["A", 12, "B", 4, "A", 6, "B", 1]。解压缩是把列表中的每个值重复指定的次数。

编写一个递归函数，来解压缩一个运行长度编码的列表。递归函数使用一个运行长度压缩列表作为其唯一的参数，返回解压缩的列表作为它的唯一结果。创建一个 main 程序，显示一个运行长度编码的列表和解码的结果。

练习 186：运行长度编码

(有解答，38 行)

编写一个递归函数，实现练习 185 中描述的运行长度压缩技术。函数以列表或字符串作为唯一参数，返回运行长度压缩列表作为它的唯一结果。包含一个 main 程序，它从用户处读取一个字符串，进行压缩，并显示运行长度编码的结果。

提示：可在递归函数体中包含一个循环。

第II部分

答 案

第9章 "编程概论"练习答案

练习1答案：邮寄地址

```
##
# Display a person's complete mailing address.
#
print("Ben Stephenson")
print("Department of Computer Science")
print("University of Calgary")
print("2500 University Drive NW")
print("Calgary, Alberta T2N 1N4")
print("Canada")
```

练习3答案：房间的面积

```
##
# Compute the area of a room.
#

# Read the dimensions from the user
length = float(input("Enter the length of the room in feet: "))
width = float(input("Enter the width of the room in feet: "))

# Compute the area of the room
area = length * width
```

> float 函数用于将用户的输入转换为数字。

> 在 Python 中，通过*操作符来执行乘法。

```
# Display the result
print("The area of the room is", area, "square feet")
```

练习 4 答案：农场的面积

```
##
# Compute the area of a field, reporting the result in acres.
#
SQFT_PER_ACRE = 43560

# Read the dimensions from the user
length = float(input("Enter the length of the field in feet: "))
width = float(input("Enter the width of the field in feet: "))

# Compute the area in acres
acres = length * width / SQFT_PER_ACRE

# Display the result
print("The area of the field is", acres, "acres"))
```

练习 5 答案：瓶子押金

```
##
# Compute the refund amount for a collection of bottles.
#
LESS_DEPOSIT = 0.10
MORE_DEPOSIT = 0.25

# Read the number of containers of each size from the user
less = int(input("How many containers 1 litre or less? "))
more = int(input("How many containers more than 1 litre? "))

# Compute the refund amount
refund = less * LESS_DEPOSIT + more * MORE_DEPOSIT

# Display the result
print("Your total refund will be $%.2f." % refund)
```

%.2f 格式说明符指出值应该格式化为小数点右边有 2 位的浮点数。

练习 6 答案：征税和小费

```
##
# Compute the tax and tip for a restaurant meal.
#
TAX_RATE = 0.05
TIP_RATE = 0.18
```

第 9 章 "编程概论"练习答案

本地税率是 5%。在 Python 中,将 5%和 18%分别表示为 0.05 和 0.18。

```
# Read the cost of the meal from the user
cost = float(input("Enter the cost of the meal: "))

# Compute the tax and the tip
tax = cost * TAX_RATE
tip = cost * TIP_RATE
total = cost + tax + tip

# Display the result
print("The tax is %.2f and the tip is %.2f, making the", \
      "total %.2f" % (tax, tip, total))
```

行末的\称为行延续符。它告诉 Python,语句在下一行继续。不要在\字符后包含任何空格或制表符。

练习 7 答案:前 *n* 个正整数的和

```
##
# Compute the sum of the first n positive integers.
#

# Read the value of n from the user
n = int(input("Enter a positive integer: "))

# Compute the sum
sm = n * (n + 1) / 2
```

Python 包含一个名为 sum()的内置函数。因此,为变量使用另一个不同的名称。

```
# Display the result
print("The sum of the first", n, "positive integers is", sm)
```

练习 10 答案:算术

```
##
# Demonstrate Python's mathematical operators and its math module.
#
from math import log10
```

在调用 log10()函数之前,必须从 math 模块中导入它。导入语句通常出现在文件顶部。

```
# Read the input values from the user
a = int(input("Enter the value of a: "))
b = int(input("Enter the value of b: "))
```

```
# Compute and display the sum, difference, product, quotient and remainder
print(a, "+", b, "is", a + b)
print(a, "-", b, "is", a - b)
print(a, "*", b, "is", a * b)
print(a, "/", b, "is", a / b)
print(a, "%", b, "is", a % b)
```

> 余数使用%运算符计算。

```
# Compute the logarithm and the power
print("The base 10 logarithm of", a, "is", log10(a))
print(a, "^", b, "is", a**b)
```

练习 13 答案：找零钱

```
##
# Compute the minimum collection of coins needed to represent a number of cents.
#
CENTS_PER_TOONIE = 200
CENTS_PER_LOONIE = 100
CENTS_PER_QUARTER = 25
CENTS_PER_DIME = 10
CENTS_PER_NICKEL = 5

# Read the number of cents from the user
cents = int(input("Enter the number of cents: "))

# Determine the number of toonies by performing an integer division by 200. Then compute
# the amount of change that still needs to be considered by computing the remainder after
# dividing by 200.
print(" ", cents // CENTS_PER_TOONIE, "toonies")
cents = cents % CENTS_PER_TOONIE
```

> 使用//操作符来执行求商除法，它通过向下舍入来确保除法的结果是整数。

```
# Repeat the process for loonies, quarters, dimes, and nickels
print(" ", cents // CENTS_PER_LOONIE, "loonies")
cents = cents % CENTS_PER_LOONIE

print(" ", cents // CENTS_PER_QUARTER, "quarters")
cents = cents % CENTS_PER_QUARTER

print(" ", cents // CENTS_PER_DIME, "dimes")
cents = cents % CENTS_PER_DIME

print(" ", cents // CENTS_PER_NICKEL, "nickels")
cents = cents % CENTS_PER_NICKEL

# Display the number of pennies
```

```
    print(" ", cents, "pennies")
```

练习 14 答案:身高单位

```
##
# Convert a height in feet and inches to centimeters.
#
IN_PER_FT = 12
CM_PER_IN = 2.54

# Read the height from the user
print("Enter your height:")
feet = int(input(" Number of feet: "))
inches = int(input(" Number of inches: "))

# Compute the equivalent number of centimeters
cm = (feet * IN_PER_FT + inches) * CM_PER_IN

# Display the result
print("Your height in centimeters is:", cm)
```

练习 17 答案:比热容

```
##
# Compute the amount of energy needed to heat a volume of water, and the cost of doing so.
#

# Define constants for the specific heat capacity of water and the price of electricity
WATER_HEAT_CAPACITY = 4.186
ELECTRICITY_PRICE = 8.9
J_TO_KWH = 2.777e-7
```

Python 允许用科学计数法来表示数字,方法是把系数放在 e 的左边,指数放在 e 的右边。结果,$2.777*10^{-7}$ 写为 2.777e-7。

```
# Read the volume and temperature increase from the user
volume = float(input("Amount of water in milliliters: "))
d_temp = float(input("Temperature increase (degrees Celsius): "))
```

因为水的密度是每毫升 1 克,而毫升可以交替使用。提示用户输入毫升值,使程序更容易使用,因为大多数人考虑的是咖啡杯中水的体积,而不是它的质量。

```
# Compute the energy in Joules
q = volume * d_temp * WATER_HEAT_CAPACITY
```

```
# Display the result in Joules
print("That will require %d Joules of energy." % q)

# Compute the cost
kwh = q * J_TO_KWH
cost = kwh * ELECTRICITY_PRICE

# Display the cost
print("That much energy will cost %.2f cents." % cost)
```

练习 19 答案：自由落体

```
##
# Compute the speed of an object when it hits the ground after being dropped.
#
from math import sqrt

# Define a constant for the acceleration due to gravity in m/s**2
GRAVITY = 9.8

# Read the height from which the object is dropped
d = float(input("Height (in meters): "))

# Compute the final velocity
vf = sqrt(2 * GRAVITY * d)
```

因为 v_i 是 0，所以 v_i^2 项没有包含在 v_f 的计算中。

```
# Display the result
print("It will hit the ground at %.2f m/s." % vf)
```

练习 23 答案：正多边形的面积

```
##
# Compute the area of a regular polygon.
#
from math import tan, pi

# Read input from the user
s = float(input("Enter the length of each side: "))
n = int(input("Enter the number of sides: "))
```

选择将 n 转换为整数(而不是浮点数)，因为多边形的边数不能是小数。

```
# Compute the area of the polygon
area = (n * s ** 2) / (4 * tan(pi / n))
```

第 9 章 "编程概论"练习答案

```
# Display the result
print("The area of the polygon is", area)
```

练习 25 答案:时间单位

```
##
# Convert a number of seconds to days, hours, minutes and seconds.
#
SECONDS_PER_DAY = 86400
SECONDS_PER_HOUR = 3600
SECONDS_PER_MINUTE = 60

# Read the duration from the user in seconds
seconds = int(input("Enter a number of seconds: "))

# Compute the days, hours, minutes and seconds
days = seconds / SECONDS_PER_DAY
seconds = seconds % SECONDS_PER_DAY
hours = seconds / SECONDS_PER_HOUR
seconds = seconds % SECONDS_PER_HOUR

minutes = seconds / SECONDS_PER_MINUTE
seconds = seconds % SECONDS_PER_MINUTE

# Display the result with the desired formatting
print("The equivalent duration is", \
    "%d:%02d:%02d:%02d." % (days, hours, minutes, seconds))
```

%02d 格式说明符告诉 Python,使用两个数字来格式化一个整数,如有必要,可以添加一个前导 0。

练习 29 答案:风寒

```
##
# Compute the wind chill index for a given air temperature and wind speed.
#
WC_OFFSET = 13.12
WC_FACTOR1 = 0.6215
WC_FACTOR2 = -11.37
WC_FACTOR3 = 0.3965
WC_EXPONENT = 0.16
```

计算风寒需要几个由科学家和医学专家确定的数值常数。

```
# Read the air temperature and wind speed from the user
temp = float(input("Air temperature (degrees Celsius): "))
speed = float(input("Wind speed (kilometers per hour): "))

# Compute the wind chill index
```

```
wci = WC_OFFSET + \
    WC_FACTOR1 * temp + \
    WC_FACTOR2 * speed ** WC_EXPONENT + \
    WC_FACTOR3 * temp * speed ** WC_EXPONENT

# Display the result rounded to the closest integer
print("The wind chill index is", round(wci))
```

练习 33 答案：对 3 个整数排序

```
##
# Sort 3 values entered by the user into increasing order.
#

# Read the numbers from the user, naming them a, b and c
a = int(input("Enter the first number: "))
b = int(input("Enter the second number: "))
c = int(input("Enter the third number: "))
mn = min(a, b, c)          # the minimum value
mx = max(a, b, c)          # the maximum value
md = a + b + c - mn - mx   # the middle value

# Display the result
print("The numbers in sorted order are:")
print("  ", mn)
print("  ", md)
print("  ", mx)
```

> 因为 min 和 max 是 Python 中函数的名称，所以不应该对变量使用这些名称。而应使用名为 mn 和 mx 的变量分别保存最小值和最大值。

练习 34 答案：旧面包

```
##
# Compute the price of a day old bread order.
#
BREAD_PRICE = 3.49
DISCOUNT_RATE = 0.60

# Read the number of loaves from the user
num_loaves = int(input("Enter the number of day old loaves: "))

# Compute the discount and total price
regular_price = num_loaves * BREAD_PRICE
discount = regular_price * DISCOUNT_RATE
total = regular_price - discount

# Display the result
print("Regular price: %5.2f" % regular_price)
print("Discount:      %5.2f" % discount)
```

```
        print("Total:          %5.2f" % total)
```

%5.2f 格式告诉 Python，总共应该使用至少 5 个空格来显示数字，小数点的右边应该有 2 位数字。当常规价格、折扣和/或总价所需的位数不同时，这将有助于保持列的对齐。

第 10 章 "决策"练习答案

练习 35 答案：偶数还是奇数？

```
##
# Determine and display whether an integer entered by the user is even or odd.
#

# Read the integer from the user
num = int(input("Enter an integer: "))

# Determine whether it is even or odd by using the
# modulus (remainder) operator
if num % 2 == 1:
    print(num, "is odd.")
else:
    print(num, "is even.")
```

> 偶数除以 2，余数总是等于 0。奇数除以 2，余数总是 1。

练习 37 答案：元音或辅音

```
##
# Determine if a letter is a vowel or a consonant.
#

# Read a letter from the user
letter = input("Enter a letter: ")
```

```
# Classify the letter and report the result
if letter == "a" or letter == "e" or \
   letter == "i" or letter == "o" or \
   letter == "u":
    print("It's a vowel.")
elif letter == "y":
    print("Sometimes it's a vowel... Sometimes it's a consonant.")
else:
    print("It's a consonant.")
```

> 这个版本的程序只适用于小写字母。可以通过包含遵循相同模式的附加比较来添加对大写字母的支持。

练习 38 答案：形状的命名

```
##
# Report the name of a shape from its number of sides.
#

# Read the number of sides from the user
nsides = int(input("Enter the number of sides: "))

# Determine the name, leaving it empty if an unsupported number of sides was entered
name = ""
if nsides == 3:
    name = "triangle"
elif nsides == 4:
    name = "quadrilateral"
elif nsides == 5:
    name = "pentagon"
elif nsides == 6:
    name = "hexagon"
elif nsides == 7:
    name = "heptagon"
elif nsides == 8:
    name = "octagon"
elif nsides == 9:
    name = "nonagon"
elif nsides == 10:
    name = "decagon"

# Display an error message or the name of the polygon
if name == "":
    print("That number of sides is not supported by this program.")
else:
    print("That's a", name)
```

> 空字符串用作标记值。如果用户输入的边数超出了支持的范围，则名称为空，从而导致在程序的后面显示一条错误消息。

第 10 章 "决策"练习答案

练习 39 答案：月名到天数

```
##
# Display the number of days in a month.
#

# Read the month name from the user
month = input("Enter the name of a month: ")

# Compute the number of days in the month
days = 31
```

首先假设天数是 31。然后根据需要更新天数。

```
if month == "April" or month == "June" or \
   month == "September" or month == "November":
    days = 30
elif month == "February":
    days = "28 or 29"
```

当月份是二月时，分配给 days 的值是一个字符串。这允许指出，二月可以有 28 天或 29 天。

```
# Display the result
print(month, "has", days, "days in it.")
```

练习 41 答案：三角形的分类

```
##
# Classify a triangle based on the lengths of its sides.
#

# Read the side lengths from the user
side1 = float(input("Enter the length of side 1: "))
side2 = float(input("Enter the length of side 2: "))
side3 = float(input("Enter the length of side 3: "))

# Determine the triangle's type
if side1 == side2 and side2 == side3:
    tri_type = "equilateral"
elif side1 == side2 or side2 == side3 or \
     side3 == side1:
    tri_type = "isosceles"
else:
    tri_type = "scalene"

# Display the triangle's type
print("That's a", tri_type, "triangle")
```

还可检查 side1 是否等于 side3，这是等边三角形的条件之一。但是，没有必要进行比较，因为==操作符是可传递的。

练习 42 答案：音符的频率

```
##
# Convert the name of a note to its frequency.
#
C4_FREQ = 261.63
D4_FREQ = 293.66
E4_FREQ = 329.63
F4_FREQ = 349.23
G4_FREQ = 392.00
A4_FREQ = 440.00
B4_FREQ = 493.88

# Read the note name from the user
name = input("Enter the two character note name, such as C4: ")

# Store the note and its octave in separate variables
note = name[0]
octave = int(name[1])

# Get the frequency of the note, assuming it is in the fourth octave
if note == "C":
    freq = C4_FREQ
elif note == "D":
    freq = D4_FREQ
elif note == "E":
    freq = E4_FREQ
elif note == "F":
    freq = F4_FREQ
elif note == "G":
    freq = G4_FREQ
elif note == "A":
    freq = A4_FREQ
elif note == "B":
    freq = B4_FREQ

# Now adjust the frequency to bring it into the correct octave
freq = freq / 2 ** (4 - octave)

# Display the result
print("The frequency of", name, "is", freq)
```

练习 43 答案：音符频率的逆转换

```
##
# Read a frequency from the user and display the note (if any) that it corresponds to.
#
```

```
C4_FREQ = 261.63
D4_FREQ = 293.66
E4_FREQ = 329.63
F4_FREQ = 349.23
G4_FREQ = 392.00
A4_FREQ = 440.00
B4_FREQ = 493.88
LIMIT = 1

# Read the frequency from the user
freq = float(input("Enter a frequency (Hz): "))

# Determine the note that corresponds to the entered frequency. Set note equal to the empty
# string if there isn't a match.
if freq >= C4_FREQ - LIMIT and freq <= C4_FREQ + LIMIT:
  note = "C4"
elif freq >= D4_FREQ - LIMIT and freq <= D4_FREQ + LIMIT:
  note = "D4"
elif freq >= E4_FREQ - LIMIT and freq <= E4_FREQ + LIMIT:
  note = "E4"
elif freq >= F4_FREQ - LIMIT and freq <= F4_FREQ + LIMIT:
  note = "F4"
elif freq >= G4_FREQ - LIMIT and freq <= G4_FREQ + LIMIT:
  note = "G4"
elif freq >= A4_FREQ - LIMIT and freq <= A4_FREQ + LIMIT:
  note = "A4"
elif freq >= B4_FREQ - LIMIT and freq <= B4_FREQ + LIMIT:
  note = "B4"
else:
  note = ""

# Display the result, or an appropriate error message
if note == "":
  print("There is no note that corresponds to that frequency.")
else:
  print("That frequency is", note)
```

练习 47 答案：季节的确定日期

```
##
# Determine and display the season associated with a date.
#

# Read the date from the user
month = input("Enter the name of the month: ")
day = int(input("Enter the day number: "))
```

这个季节识别问题的解使用了几个 elif 部件，使情况尽可能简单。解决此问题的另一种方法是通过使条件更复杂，来最小化 elif 部件的数量。

```python
# Determine the season
if month == "January" or month == "February":
  season = "Winter"
elif month == "March":
  if day < 20:
    season = "Winter"
  else:
    season = "Spring"
elif month == "April" or month == "May":
  season = "Spring"
elif month == "June":
  if day < 21:
    season = "Spring"
  else:
    season = "Summer"
elif month == "July" or month == "August":
  season = "Summer"
elif month == "September":
  if day < 22:
    season = "Summer"
  else:
    season = "Fall"
elif month == "October" or month == "November":
  season = "Fall"
elif month == "December":
  if day < 21:
    season = "Fall"
  else:
    season = "Winter"

# Display the result
print(month, day, "is in", season)
```

练习 49 答案：十二生肖

```python
##
# Determine the animal associated with a year according to the Chinese zodiac.
#

# Read a year from the user
year = int(input("Enter a year: "))

# Determine the animal associated with that year
```

第 10 章 "决策"练习答案

```
    if year % 12 == 8:
        animal = "Dragon"
    elif year % 12 == 9:
        animal = "Snake"
    elif year % 12 == 10:
        animal = "Horse"
    elif year % 12 == 11:
        animal = "Sheep"
    elif year % 12 == 0:
        animal = "Monkey"
    elif year % 12 == 1:
        animal = "Rooster"
    elif year % 12 == 2:
        animal = "Dog"
    elif year % 12 == 3:
        animal = "Pig"
    elif year % 12 == 4:
        animal = "Rat"
    elif year % 12 == 5:
        animal = "Ox"
    elif year % 12 == 6:
        animal = "Tiger"
    elif year % 12 == 7:
        animal = "Hare"

    # Report the result
    print("%d is the year of the %s." % (year, animal))
```

当格式化多个项时，所有的值都放在%运算符右侧的圆括号内。

练习 52 答案：字母等级到绩点

```
##
# Convert from a letter grade to a number of grade points.
#
A       = 4.0
A_MINUS = 3.7
B_PLUS  = 3.3
B       = 3.0
B_MINUS = 2.7
C_PLUS  = 2.3
C       = 2.0
C_MINUS = 1.7
D_PLUS  = 1.3
D       = 1.0
F       = 0
```

```
INVALID = -1

# Read the letter grade from the user
letter = input("Enter a letter grade: ")
letter = letter.upper()
```

语句 letter = letter.upper()将用户输入的任何小写字母转换成大写字母,将结果存储回同一变量中。包含这个语句允许程序使用小写字母,而不需要在 if 和 elif 部分的条件中包含它们。

```
# Convert from a letter grade to a number of grade points using -1 grade points as a sentinel
# value indicating invalid input
if letter == "A+" or letter == "A":
  gp = A
elif letter == "A-":
  gp = A_MINUS
elif letter == "B+":
  gp = B_PLUS
elif letter == "B":
  gp = B
elif letter == "B-":
  gp = B_MINUS
elif letter == "C+":
  gp = C_PLUS
elif letter == "C":
  gp = C
elif letter == "C-":
  gp = C_MINUS
elif letter == "D+":
  gp = D_PLUS
elif letter == "D":
  gp = D
elif letter == "F":
  gp = F
else:
  gp = INVALID

# Report the result
if gp == INVALID:
  print("That wasn't a valid letter grade.")
else:
  print("A(n)", letter, "is equal to", gp, "grade points.")
```

第 10 章 "决策"练习答案

练习 54 答案：评估员工

```
##
# Report whether an employee's performance is unacceptable, acceptable or meritorious
# based on the rating entered by the user.
#
RAISE_FACTOR = 2400.00
UNACCEPTABLE = 0
ACCEPTABLE = 0.4
MERITORIOUS = 0.6

# Read the rating from the user
rating = float(input("Enter the rating: "))

# Classify the performance
if rating == UNACCEPTABLE:
  performance = "Unacceptable"
elif rating == ACCEPTABLE:
  performance = "Acceptable"
elif rating >= MERITORIOUS:
  performance = "Meritorious"
else:
  performance = ""

# Report the result
if performance == "":
  print("That wasn't a valid rating.")
else:
  print("Based on that rating, your performance is %s." % \
        performance)
print("You will receive a raise of $%.2f." % \
      (rating * RAISE_FACTOR))
```

> 最后一行中围绕 rating *RAISE_FACTOR 的括号是必要的，因为%和*的优先级相同。包含括号迫使 Python 在格式化结果之前执行数学计算。

练习 58 答案：这是闰年吗？

```
##
# Determine whether or not a year is a leap year.
#

# Read the year from the user
year = int(input("Enter a year: "))

# Determine if it is a leap year
```

```
if year % 400 == 0:
  isLeapYear = True
elif year % 100 == 0:
  isLeapYear = False
elif year % 4 == 0:
  isLeapYear = True
else:
  isLeapYear = False

# Display the result
if isLeapYear:
  print(year, "is a leap year.")
else:
  print(year, "is not a leap year.")
```

练习 61 答案：车牌有效吗？

```
## Determine whether or not a license plate is valid. A valid license plate either consists of:
#   1) 3 letters followed by 3 numbers, or
#   2) 4 numbers followed by 3 numbers

# Read the plate from the user
plate = input("Enter the license plate: ")

# Check the status of the plate and display it. It is necessary to check each of the 6 characters
# for an older style plate, or each of the 7 characters for a newer style plate.
if len(plate) == 6 and \
   plate[0] >= "A" and plate[0] <= "Z" and \
   plate[1] >= "A" and plate[1] <= "Z" and \
   plate[2] >= "A" and plate[2] <= "Z" and \
   plate[3] >= "0" and plate[3] <= "9" and \
   plate[4] >= "0" and plate[4] <= "9" and \
   plate[5] >= "0" and plate[5] <= "9":
  print("The plate is a valid older style plate.")
elif len(plate) == 7 and \
   plate[0] >= "0" and plate[0] <= "9" and \
   plate[1] >= "0" and plate[1] <= "9" and \
   plate[2] >= "0" and plate[2] <= "9" and \
   plate[3] >= "0" and plate[3] <= "9" and \
   plate[4] >= "A" and plate[4] <= "Z" and \
   plate[5] >= "A" and plate[5] <= "Z" and \
   plate[6] >= "A" and plate[6] <= "Z":
  print("The plate is a valid newer style plate.")
else:
  print("The plate is not valid.")
```

第 10 章 "决策"练习答案

练习 62 答案：轮盘赌

```python
##
# Display the bets that pay out in a roulette simulation.
#
from random import randrange

# Simulate spinning the wheel, using 37 to represent 00
value = randrange(0, 38)
if value == 37:
    print("The spin resulted in 00...")
else:
    print("The spin resulted in %d..." % value)

# Display the payout for a single number
if value == 37:
    print("Pay 00")
else:
    print("Pay", value)

# Display the color payout
# The first line in the condition checks for 1, 3, 5, 7 and 9
# The second line in the condition checks for 12, 14, 16 and 18
# The third line in the condition checks for 19, 21, 23, 25 and 27
# The fourth line in the condition checks for 30, 32, 34 and 36
if value % 2 == 1 and value >= 1 and value <= 9 or \
   value % 2 == 0 and value >= 12 and value <= 18 or \
   value % 2 == 1 and value >= 19 and value <= 27 or \
   value % 2 == 0 and value >= 30 and value <= 36:
    print("Pay Red")
elif value == 0 or value == 37:
    pass
else:
    print("Pay Black")

# Display the odd vs. even payout
if value >= 1 and value <= 36:
    if value % 2 == 1:
        print("Pay Odd")
    else:
        print("Pay Even")

# Display the lower numbers vs. upper numbers payout
if value >= 1 and value <= 18:
    print("Pay 1 to 18")
elif value >= 19 and value <= 36:
    print("Pay 19 to 36")
```

> if、elif 或 else 的主体必须包含至少一条语句。Python 包含 pass 关键字，当需要一条语句但不需要执行任何工作时，可以使用这个关键字。

第11章 "循环"练习答案

练习 66 答案：不要再花钱了

```
##
# Compute the total due when several items are purchased. The amount payable for cash
# transactions is rounded to the closest nickel because pennies have been phased out in Canada.
#
PENNIES_PER_NICKEL = 5
NICKEL = 0.05
```

> 虽然一美分的分币数不太可能发生变化，但可能需要在未来某个时候更新程序，使其精确到最接近的分币值。使用常量更容易在需要时执行更新。

```
# Track the total cost for all of the items
total = 0.00

# Read the price of the first item as a string
line = input("Enter the price of the item (blank to quit): ")

# Continue reading items until a blank line is entered
while line != "":
    # Add the cost of the item to the total (after converting it to a floating-point number)
    total = total + float(line)

    # Read the cost of the next item
    line = input("Enter the price of the item (blank to quit): ")
```

```
# Display the exact total payable
print("The exact amount payable is %.02f" % total)

# Compute the number of pennies that would be left if the total was paid using nickels
rounding_indicator = total * 100 % PENNIES_PER_NICKEL
if rounding_indicator < PENNIES_PER_NICKEL / 2:
    # If the number of pennies left is less than 2.5 then we round down by subtracting that
    # number of pennies from the total
    cash_total = total - rounding_indicator / 100
else:
    # Otherwise we add a nickel and then subtract that number of pennies
    cash_total = total + NICKEL - rounding_indicator / 100

# Display amount due when paying with cash
print("The cash amount payable is %.02f" % cash_total)
```

练习 67 答案：计算多边形的周长

```
##
# Compute the perimeter of a polygon constructed from points entered by the user. A blank line
# will be entered for the x-coordinate to indicate that all of the points have been entered.
#
from math import sqrt

# Store the perimeter of the polygon
perimeter = 0

# Read the coordinate of the first point
first_x = float(input("Enter the first x-coordinate: "))
first_y = float(input("Enter the first y-coordinate: "))

# Provide initial values for prev x and prev y
prev_x = first_x
prev_y = first_y

# Read the remaining coordinates
line = input("Enter the next x-coordinate (blank to quit): ")
while line != "":
    # Convert the x-coordinate to a number and read the y coordinate
    x = float(line)
    y = float(input("Enter the next y-coordinate: "))

    # Compute the distance to the previous point and add it to the perimeter
    dist = sqrt((prev_x - x) ** 2 + (prev_y - y) ** 2)
    perimeter = perimeter + dist

    # Set up prev x and prev y for the next loop iteration
    prev_x = x
```

> 这些点之间的距离是用勾股定理计算的。

第 11 章 "循环"练习答案

```
    prev_y = y

    # Read the next x-coordinate
    line = input("Enter the next x-coordinate (blank to quit): ")

# Compute the distance from the last point to the first point and add it to the perimeter
dist = sqrt((first_x - x) ** 2 + (first_y - y) ** 2)
perimeter = perimeter + dist

# Display the result
print("The perimeter of that polygon is", perimeter)
```

练习 69 答案：票价

```
##
# Compute the admission price for a group visiting the zoo.
#

# Store the admission prices as constants
BABY_PRICE = 0.00
CHILD_PRICE = 14.00
ADULT_PRICE = 23.00
SENIOR_PRICE = 18.00

# Store the age limits as constants
BABY_LIMIT = 2
CHILD_LIMIT = 12
ADULT_LIMIT = 64

# Create a variable to hold the total admission cost for all guests
total = 0

# Display the total due for the group, formatted using two decimal places
print("The total for that group is $%.2f" % total)

# Keep on reading ages until the user enters a blank line
line = input("Enter the age of the guest (blank to finish): ")
while line != "":
  age = int(line)

  # Add the correct amount to the total
  if age <= BABY_LIMIT:
    total = total + BABY_PRICE
  elif age <= CHILD_LIMIT:
    total = total + CHILD_PRICE
  elif age <= ADULT_LIMIT:
    total = total + ADULT_PRICE
  else:
```

> 按照目前的门票价格，if-elif-else 语句的第一部分可能会被删除。然而，包括它使程序更容易更新，以便将来对婴儿收费。

```
        total = total + SENIOR_PRICE

    # Read the next age from the user
    line = input("Enter the age of the guest (blank to finish): ")

# Display the total due for the group, formatted using two decimal places
print("The total for that group is $%.2f" % total)
```

练习 70 答案：奇偶校验位

```
##
# Compute the parity bit using even parity for sets of 8 bits entered by the user.
#

# Read the first line of input
line = input("Enter 8 bits: ")

# Continue looping until a blank line is entered
while line != "":
  # Ensure that the line has a total of 8 zeros and ones and exactly 8 characters
  if line.count("0") + line.count("1") != 8 or len(line) != 8:
    # Display an appropriate error message
    print("That wasn't 8 bits... Try again.")
  else:
    # Count the number of ones
    ones = line.count("1")
```

count 方法返回参数在调用字符串中的出现次数。

```
    # Display the parity bit
    if ones % 2 == 0:
      print("The parity bit should be 0.")
    else:
      print("The parity bit should be 1.")

  # Read the next line of input
  line = input("Enter 8 bits: ")
```

练习 73 答案：凯撒密码

```
##
# Implement a Caesar cipher that shifts all of the letters in a message by an amount provided
# by the user. Use a negative shift value to decode a message.
#

# Read the message and shift amount from the user
message = input("Enter the message: ")
```

```
shift = int(input("Enter the shift value: "))
# Process each character to construct the encrypted (or decrypted) message
new_message = ""
for ch in message:
  if ch >= "a" and ch <= "z":
    # Process a lowercase letter by determining its
    # position in the alphabet (0 - 25), computing its
    # new position, and adding it to the new message
    pos = ord(ch) - ord("a")
    pos = (pos + shift) % 26
    new_char = chr(pos + ord("a"))
    new_message = new_message + new_char
  elif ch >= "A" and ch <= "Z":
    # Process an uppercase letter by determining its position in the alphabet (0 - 25),
    # computing its new position, and adding it to the new message
    pos = ord(ch) - ord("A")
    pos = (pos + shift) % 26
    new_char = chr(pos + ord("A"))
    new_message = new_message + new_char
  else:
    # If the character is not a letter then copy it into the new message
    new_message = new_message + ch

# Display the shifted message
print("The shifted message is", new_message)
```

> ord 函数将字符转换为其在 ASCII 表中的整数位置。chr 函数返回位置参数在 ASCII 表中的对应字符。

练习 75 答案：字符串是回文吗？

```
##
# Determine whether or not a string entered by the user is a palindrome.
#

# Read the string from the user
line = input("Enter a string: ")

# Assume that it is a palindrome until we can prove otherwise
is_palindrome = True

# Check the characters, starting from the ends. Continue until the middle is reached or we have
# determined that the string is not a palindrome.
i = 0
while i < len(line) / 2 and is_palindrome:
  # If the characters do not match then mark that the string is not a palindrome
  if line[i] != line[len(line) - i - 1]:
    is_palindrome = False
```

```
    # Move to the next character
    i = i + 1

# Display a meaningful output message
if is_palindrome:
    print(line, "is a palindrome")
else:
    print(line, "is not a palindrome")
```

练习 77 答案：乘法表

```
##
# Display a multiplication table for 1 times 1 through 10 times 10.
#
MIN = 1
MAX = 10

# Display the top row of labels
print("  ", end="")
for i in range(MIN, MAX + 1):
    print("%4d" % i, end="")
print()

# Display the table
for i in range(MIN, MAX + 1):
    print("%4d" % i, end="")
    for j in range(MIN, MAX + 1):
        print("%4d" % (i * j), end="")
print()
```

包含 end=""作为要打印的最后一个参数可防止它在显示值之后向下移动到下一行。

练习 79 答案：最大公约数

```
##
# Compute the greatest common divisor of two positive integers using a while loop.
#

# Read two positive integers from the user
n = int(input("Enter a positive integer: "))
m = int(input("Enter a positive integer: "))

# Initialize d to the smaller of n and m
d = min(n, m)

# Use a while loop to find the greatest common divisor of n and m
while n % d != 0 or m % d != 0:
    d = d - 1
```

第 11 章 "循环"练习答案

```
# Report the result
print("The greatest common divisor of", n, "and", m, "is", d)
```

练习 82 答案：十进制转换到二进制

```
##
# Convert a number from decimal (base 10) to binary (base 2).
#
NEW_BASE = 2

# Read the number to convert from the user
num = int(input("Enter a non-negative integer: "))
# Generate the binary representation of num, storing it in result
result = ""
q = num
```

> 为这个问题提供的算法是使用一个 repeat-until 循环来表示的。然而，这种类型的循环在 Python 中不可用。因此，必须对算法进行调整，以便使用 while 循环生成相同的结果。这是通过复制循环体并将其置于 while 循环之前来实现的。

```
# Perform the body of the loop once
r = q % NEW_BASE
result = str(r) + result
q = q // NEW_BASE

# Keep on looping until q is 0
while q > 0:
    r = q % NEW_BASE
    result = str(r) + result
    q = q // NEW_BASE

# Display the result
print(num, "in decimal is", result, "in binary.")
```

练习 83 答案：最大整数

```
##
# Find the maximum of 100 random integers and count the number of times the maximum value
# is updated during the process.
#
from random import randrange


NUM_ITEMS = 100

# Generate the first number and display it
```

```
mx_value = randrange(1, NUM_ITEMS + 1)
print(mx_value)

# Count the number of times the maximum value is updated
num_updates = 0

# For each of the remaining numbers
for i in range(1, NUM_ITEMS):
    # Generate a new random number
    current = randrange(1, NUM_ITEMS + 1)

    # If the generated number is the largest one we have seen so far
    if current > mx_value:
        # Update the maximum and count the update
        mx_value = current
        num_updates = num_updates + 1
        # Display the number, noting that an update occurred
        print(current, "<== Update")
    else:
        # Display the number
        print(current)

# Display the other results
print("The maximum value found was", mx_value)
print("The maximum value was updated", num_updates, "times")
```

第 12 章 "函数"练习答案

练习 88 答案：三个值的中位数

```
##
# Compute and display the median of three values entered by the user. This program includes
# two implementations of the median function that demonstrate different techniques for
# computing the median of three values.
#

## Compute the median of three values using if statements
# @param a the first value
# @param b the second value
# @param c the third value
# @return the median of values a, b and c
#
def median(a, b, c):
  if a < b and b < c or a > b and b > c:
    return b
  if b < a and a < c or b > a and a > c:
    return a
  if c < a and b < c or c > a and b > c:
    return c

## Compute the median of three values using the min and max functions and a little bit of
# arithmetic
```

> 编写的每个函数都应该以注释开头。以 @param 开头的行用于描述函数的参数。函数所返回的值由一个带有 @return 的行表示。

```
# @param a the first value
# @param b the second value
# @param c the third value
# @return the median of values a, b and c
#
def alternateMedian(a, b, c):
    return a + b + c - min(a, b, c) - max(a, b, c)
```

> 三个值的中位数是这三个值之和，减去最小值，再减去最大值。

```
x# Display the median of 3 values entered by the user
def main():
    x = float(input("Enter the first value: "))
    y = float(input("Enter the second value: "))
    z = float(input("Enter the third value: "))
    print("The median value is:", median(x, y, z))
    print("Using the alternative method, it is:", \
          alternateMedian(x, y, z))

# Call the main function
main()
```

练习 90 答案：圣诞节的 12 天

```
##
# Display the complete lyrics for the song The Twelve Days of Christmas.
#
from int_ordinal import intToOrdinal
```

> 为前一个练习编写的函数导入这个程序中，这样从整数转换为序数的代码就不必在这里重复了。

```
## Display one verse of The Twelve Days of Christmas
# @param n the verse to display
# @return (None)
def displayVerse(n):
    print("On the", intToOrdinal(n), "day of Christmas")
    print("my true love sent to me:")

    if n >= 12:
        print("Twelve drummers drumming,")
    if n >= 11:
        print("Eleven pipers piping,")
    if n >= 10:
        print("Ten lords a-leaping,")
    if n >= 9:
        print("Nine ladies dancing,")
    if n >= 8:
```

```
    print("Eight maids a-milking,")
  if n >= 7:
    print("Seven swans a-swimming,")
  if n >= 6:
    print("Six geese a-laying,")
  if n >= 5:
    print("Five golden rings,")
  if n >= 4:
    print("Four calling birds,")
  if n >= 3:
    print("Three French hens,")
  if n >= 2:
    print("Two turtle doves,")
  if n == 1:
    print("A", end=" ")
  else:
    print("And a", end=" ")
  print("partridge in a pear tree.")
  print()

# Display all 12 verses of the song
def main():
  for verse in range(1, 13):
    displayVerse(verse)

# Call the main function
main()
```

练习 93 答案：在终端窗口中居中显示一个字符串

```
##
# Center a string of characters within a certain width.
#
WIDTH = 80

## Create a new string that will be centered within a given width when it is printed.
# @param s the string that will be centered
# @param width the width in which the string will be centered
# @return a new copy of s that contains the leading spaces needed to center s
def center(s, width):
  # If the string is too long to center, then the original string is returned
  if width < len(s):
    return s

  # Compute the number of spaces needed and generate the result
  spaces = (width - len(s)) // 2
  result = " " * spaces + s
```

使用//运算符将除法的结果截断为整数。不能使用/运算符，因为它返回浮点结果，但字符串只能复制整数次。

```
    return result

# Demonstrate the center function
def main():
    print(center("A Famous Story", WIDTH))
    print(center("by:", WIDTH))
    print(center("Someone Famous", WIDTH))
    print()
    print("Once upon a time...")

# Call the main function
main()
```

练习 95 答案：大写

```
##
# Improve the capitalization of a string.
#

## Capitalize the appropriate characters in a string
# @param s the string that needs capitalization
# @return a new string with the capitalization improved
def capitalize(s):
    # Create a new copy of the string to return as the function's result
    result = s

    # Capitalize the first non-space character in the string
    pos = 0
    while pos < len(s) and result[pos] == ' ':
        pos = pos + 1

    if pos < len(s):
        # Replace the character with its uppercase version without changing any other characters
        result = result[0 : pos] + result[pos].upper() + \
                 result[pos + 1 : len(result)]
```

在方括号内使用冒号检索字符串的一部分。检索到的字符从冒号左边的位置开始，一直到冒号右边的位置(不含)。

```
    # Capitalize the first letter that follows a ".", "!" or "?"
    pos = 0
    while pos < len(s):
        if result[pos] == "." or result[pos] == "!" or \
```

```
          result[pos] == "?":
    # Move past the ".", "!" or "?"
    pos = pos + 1

    # Move past any spaces
    while pos < len(s) and result[pos] == " ":
      pos = pos + 1

    # If we haven't reached the end of the string then replace the current character
    # with its uppercase equivalent
    if pos < len(s):
      result = result[0 : pos] + \
               result[pos].upper() + \
               result[pos + 1 : len(result)]

  # Move to the next character
  pos = pos + 1

# Capitalize i when it is preceded by a space and followed by a space, period, exclamation
# mark, question mark or apostrophe
pos = 1
while pos < len(s) - 1:
  if result[pos - 1] == " " and result[pos] == "i" and \
     (result[pos + 1] == " " or result[pos + 1] == "." or \
      result[pos + 1] == "!" or result[pos + 1] == "?" or \
      result[pos + 1] == "'"):
    # Replace the i with an I without changing any other characters
    result = result[0 : pos] + "I" + \
             result[pos + 1 : len(result)]
  pos = pos + 1
  return result

# Demonstrate the capitalize function
def main():
  s = input("Enter some text: ")
  capitalized = capitalize(s)
  print("It is capitalized as:", capitalized)

# Call the main function
main()
```

练习 96 答案：字符串是否表示整数？

```
##
# Determine whether or not a string entered by the user is an integer.
#
## Determine if a string contains a valid representation of an integer
```

```
# @param s the string to check
# @return True if s represents an integer. False otherwise.
#
def isInteger(s):
  # Remove whitespace from the beginning and end of the string
  s = s.strip()

  # Determine if the remaining characters form a valid integer
  if (s[0] == "+" or s[0] == "-") and s[1:].isdigit():
    return True
  if s.isdigit():
    return True
  return False
```

> 当且仅当字符串长度至少为一个字符且字符串中的所有字符都是数字时,isdigit 方法才返回 True。

```
# Demonstrate the isInteger function
def main():
  s = input("Enter a string: ")
  if isInteger(s):
    print("That string represents an integer.")
  else:
    print("That string does not represent an integer.")

# Only call the main function when this file has not been imported
if __name__ == "__main__":
  main()
```

> 当程序开始运行时,_name_变量由 Python 自动分配一个值。当 Python 直接执行文件时,它包含"_main_"。当文件导入另一个程序时它包含模块的名称。

练习 98 答案:一个数是素数吗?

```
##
# Determine if a number entered by the user is prime.
#

## Determine whether or not a number is prime
# @param n the integer to test
# @return True if the number is prime, False otherwise
def isPrime(n):
  if n <= 1:
    return False

  # Check each number from 2 up to but not including n to see if it divides evenly into n
  for i in range(2, n):
    if n % i == 0:
```

> 如果 n % i == 0,那么 n 能被 i 整除,表示 n 不是质数。

```
      return False
   return True

# Determine if a number entered by the user is prime
def main():
   value = int(input("Enter an integer: "))
   if isPrime(value):
      print(value, "is prime.")
   else:
      print(value, "is not prime.")

# Call the main function if the file has not been imported
if __name__ == "__main__":
   main()
```

练习 100 答案：随机密码

```
##
# Generate and display a random password containing between 7 and 10 characters.
#
from random import randint

SHORTEST = 7
LONGEST = 10
MIN_ASCII = 33
MAX_ASCII = 126
## Generate a random password
# @return a string containing a random password
def randomPassword():
   # Select a random length for the password
   randomLength = randint(SHORTEST, LONGEST)

   # Generate an appropriate number of random characters, adding each one to the end of result
   result = ""
   for i in range(randomLength):
      randomChar = chr(randint(MIN_ASCII, MAX_ASCII))
      result = result + randomChar
```

chr 函数的唯一参数是 ASCII 码。它返回一个字符串，该字符串包含结果为 ASCII 码的字符。

```
   # Return the random password
   return result

# Generate and display a random password
def main():
   print("Your random password is:", randomPassword())
```

```
# Call the main function only if the module is not imported
if __name__ == "__main__":
    main()
```

练习 102 答案：检查密码

```
##
# Check whether or not a password is good.
#

## Check whether or not a password is good. A good password is at least 8 characters and
# contains an uppercase letter, a lowercase letter and a number.
# @param password the password to check
# @return True if the password is good, False otherwise
def checkPassword(password):
    has_upper = False
    has_lower = False
    has_num = False

    # Check each character in the password and see which requirement it meets
    for ch in password:
        if ch >= "A" and ch <= "Z":
            has_upper = True
        elif ch >= "a" and ch <= "z":
            has_lower = True
        elif ch >= "0" and ch <= "9":
            has_num = True

    # If the password has all 4 properties
    if len(password) >= 8 and has_upper and has_lower and has_num:
        return True

    # The password is missing at least one property
    return False

# Demonstrate the password checking function
def main():
    p = input("Enter a password: ")
    if checkPassword(p):
        print("That's a good password.")
    else:
        print("That isn't a good password.")

# Call the main function only if the file has not been imported into another program
if __name__ == "__main__":
    main()
```

第 12 章 "函数"练习答案

练习 105 答案：任意进制之间的转换

```
##
# Convert a number from one base to another. Both the source base and the destination base
# must be between 2 and 16.
#
from hex_digit import *
```

hex_digit 模块包含了在解练习 104 时需要的函数 hex2int 和 int2hex。使用 import*将所有函数从该模块导入。

```
## Convert a number from base 10 to base new base
# @param num the base 10 number to convert
# @param new base the base to convert to
# @return the string of digits in new base
def dec2n(num, new_base):
    # Generate the representation of num in base new base, storing it in result
    result = ""
    q = num

    # Perform the body of the loop once
    r = q % new_base
    result = int2hex(r) + result
    q = q // new_base

    # Continue looping until q is 0
    while q > 0:
        r = q % new_base
        result = int2hex(r) + result
        q = q // new_base

    # Return the result
    return result
## Convert a number from base b to base 10
# @param num the base b number, stored in a string
# @param b the base of the number to convert
# @return the base 10 number
def n2dec(num, b):
    decimal = 0

    # Process each digit in the base b number
    for i in range(len(num)):
        decimal = decimal * b
        decimal = decimal + hex2int(num[i])

    # Return the result
```

基数 b 必须存储在字符串中，因为它可能包含以大于 10 的数为基数来表示数字的字母。

```
    return decimal
```

```
# Convert a number between two arbitrary bases
def main():
  # Read the base and number from the user
  from_base = int(input("Base to convert from (2-16): "))
  if from_base < 2 or from_base > 16:
    print("Only bases between 2 and 16 are supported.")
    print("Quitting...")
    quit()

  from_num = input("Sequence of digits in that base: ")

  # Convert to base 10 and display the result
  dec = n2dec(from_num, from_base)
  print("That's %d in base 10." % dec)

  # Convert to the new base and display the result
  to_base = int(input("Enter the base to convert to (2-16): "))
  if to_base < 2 or to_base > 16:
    print("Only bases between 2 and 16 are supported.")
    print("Quitting...")
    quit()

  to_num = dec2n(dec, to_base)
  print("That's %s in base %d." % (to_num, to_base))

# Call the main function
main()
```

练习 107 答案：最简分数

```
# Reduce a fraction to lowest terms.
#

## Compute the greatest common divisor of two integers
# @param n the first integer under consideration (must be non-zero)
# @param m the second integer under consideration (must be non-zero)
# @return the greatest common divisor of the integers
def gcd(n, m):
  # Initialize d to the smaller of n and m
  d = min(n, m)

# Use a while loop to find the greatest common divisor of n and m
  while n % d != 0 or m % d != 0:
    d = d - 1

  return d
```

第 12 章 "函数"练习答案

> 这个函数使用一个循环来实现目标。还有一种优雅的解决方案可使用递归找到两个整数的最大公约数。在练习 174 中探讨了递归解决方案。

```
## Reduce a fraction to lowest terms
# @param num the integer numerator of the fraction
# @param den the integer denominator of the fraction (must be non-zero)
# @return the numerator and denominator of the reduced fraction
def reduce(num, den):
  # If the numerator is 0 then the reduced fraction is 0 / 1
  if num == 0:
    return (0, 1)

  # Compute the greatest common divisor of the numerator and denominator
  g = gcd(num, den)

  # Divide both the numerator and denominator by the GCD and return the result
  return (num // g, den // g)
```

这里使用了取整除法运算符//,因此分子和分母在函数返回的结果中都是整数。

```
# Read a fraction from the user and display the equivalent lowest terms fraction
def main():
  # Read the numerator and denominator from the user
  num = int(input("Enter the numerator: "))
  den = int(input("Enter the denominator: "))

  # Compute the reduced fraction
  (n, d) = reduce(num, den)

  # Display the result
  print("%d/%d can be reduced to %d/%d." % (num, den, n, d))

# Call the main function
main()
```

练习 108 答案:减少度量单位

```
##
# Reduce an imperial measurement so that it is expressed using the largest possible units of
# measure. For example, 59 teaspoons is reduced to 1 cup, 3 tablespoons, 2 teaspoons.
#
TSP_PER_TBSP = 3
TSP_PER_CUP = 48

## Reduce an imperial measurement so that it is expressed using the largest possible
# units of measure
```

```
# @param num the number of units that need to be reduced
# @param unit the unit of measure ("cup", "tablespoon" or "teaspoon")
# @return a string representing the measurement in reduced form
def reduceMeasure(num, unit):
    # Convert the unit to lowercase
    unit = unit.lower()
```

通过调用 unit 的 lower 方法并将结果存储到相同的变量中,该单位转换为小写形式。这允许用户在指定单位时使用大写和小写字母的任意组合。

```
    # Compute the number of teaspoons that the parameters represent
    if unit == "teaspoon" or unit == "teaspoons":
        teaspoons = num
    elif unit == "tablespoon" or unit == "tablespoons":
        teaspoons = num * TSP_PER_TBSP
    elif unit == "cup" or unit == "cups":
        teaspoons = num * TSP_PER_CUP

    # Convert the number of teaspoons to the largest possible units of measure
    cups = teaspoons // TSP_PER_CUP
    teaspoons = teaspoons - cups * TSP_PER_CUP
    tablespoons = teaspoons // TSP_PER_TBSP
    teaspoons = teaspoons - tablespoons * TSP_PER_TBSP

    # Create a string to hold the result
    result = ""

    # Add the number of cups to the result string (if any)
    if cups > 0:
        result = result + str(cups) + " cup"
        # Make cup plural if there is more than one
        if cups > 1:
            result = result + "s"

    # Add the number of tablespoons to the result string (if any)
    if tablespoons > 0:
        # Include a comma if there were some cups
        if result != "":
            result = result + ", "
        result = result + str(tablespoons) + " tablespoon"
        # Make tablespoon plural if there is more than one
        if tablespoons > 1:
            result = result + "s"

    # Add the number of teaspoons to the result string (if any)
    if teaspoons > 0:
        # Include a comma if there were some cups and/or tablespoons
```

```
    if result != "":
        result = result + ", "
result = result + str(teaspoons) + " teaspoon"
# Make teaspoons plural if there is more than one
if teaspoons > 1:
    result = result + "s"

# Handle the case where the number of units was 0
if result == "":
    result = "0 teaspoons"
return result
```

> 该程序包含几个测试用例。它们运用了各种不同的组合——零、一、多次出现不同的度量单位。虽然这些测试用例相当全面，但不能保证程序没有 bug。

```
# Demonstrate the reduceMeasure function by performing several reductions
def main():
    print("59 teaspoons is %s." % reduceMeasure(59, "teaspoons"))
    print("59 tablespoons is %s." % \
          reduceMeasure(59, "tablespoons"))
    print("1 teaspoon is %s." % reduceMeasure(1, "teaspoon"))
    print("1 tablespoon is %s." % reduceMeasure(1, "tablespoon"))
    print("1 cup is %s." % reduceMeasure(1, "cup"))
    print("4 cups is %s." % reduceMeasure(4, "cups"))
    print("3 teaspoons is %s." % reduceMeasure(3, "teaspoons"))
    print("6 teaspoons is %s." % reduceMeasure(6, "teaspoons"))
    print("95 teaspoons is %s." % reduceMeasure(95, "teaspoons"))
    print("96 teaspoons is %s." % reduceMeasure(96, "teaspoons"))
    print("97 teaspoons is %s." % reduceMeasure(97, "teaspoons"))
    print("98 teaspoons is %s." % reduceMeasure(98, "teaspoons"))
    print("99 teaspoons is %s." % reduceMeasure(99, "teaspoons"))

# Call the main function
main()
```

练习 109 答案：神奇的日子

```
##
# Determine all of the magic dates in the 1900s.
#
from days_in_month import daysInMonth

## Determine whether or not a date is ''magic''
# @param day the day portion of the date
# @param month the month portion of the date
# @param year the year portion of the date
```

```
# @return True if the date is magic, False otherwise
def isMagicDate(day, month, year):
  if day * month == year % 100:
    return True

  return False
```

表达式 year % 100 的计算结果为两位数字的 year。

```
# Find and display all of the magic dates in the 1900s
def main():
  for year in range(1900, 2000):
    for month in range(1, 13):
      for day in range(1, daysInMonth(month, year) + 1):
        if isMagicDate(day, month, year):
          print("%02d/%02d/%04d is a magic date." % (day, month, year))

# Call the main function
main()
```

第13章 "列表"练习答案

练习 110 答案：排序

```
##
# Display a list of integers entered by the user in ascending order.
#
# Start with an empty list
data = []

# Read values and add them to the list until the user enters 0
num = int(input("Enter an integer (0 to quit): "))
while num != 0:
  data.append(num)
  num = int(input("Enter an integer (0 to quit): "))

# Sort the values
data.sort()
```

> 调用列表上的 sort()方法将列表中的元素重新排列为已排序的顺序。使用 sort()方法适合于此题，因为不需要保留原始列表的副本。sort()可用于创建列表的新副本，其中的元素按排序顺序排列。调用 sort()不会修改原始列表。因此，它可用于同时需要原始列表和排序列表的情况。

```
# Display the values in ascending order
```

```
print("The values, sorted into ascending order, are:")
for num in data:
  print(num)
```

练习 112 答案：删除异常值

```
##
# Remove the outliers from a data set.
#
## Remove the outliers from a list of values
# @param data the list of values to process
# @param num_outliers the number of smallest and largest values to remove
# @return a new copy of data where the values are sorted into ascending order and the
#         smallest and largest values have been removed
def removeOutliers(data, num_outliers):
  # Create a new copy of the list that is in sorted order
  retval = sorted(data)

  # Remove num outliers largest values
  for i in range(num_outliers):
    retval.pop()

  # Remove num outliers smallest values
  for i in range(num_outliers):
    retval.pop(0)

  # Return the result
  return retval

# Read data from the user, and remove the two largest and two smallest values
def main():
  # Read values from the user until a blank line is entered
  values = []
  s = input("Enter a value (blank line to quit): ")
  while s != "":
    num = float(s)
    values.append(num)
    s = input("Enter a value (blank line to quit): ")

  # Display the result or an appropriate error message
  if len(values) < 4:
    print("You didn't enter enough values.")
  else:
    print("With the outliers removed: ", \
          removeOutliers(values, 2))
    print("The original data: ", values)
```

> 最小和最大的异常值可以使用相同的循环来删除。此解使用了两个循环，以使步骤更加清晰。

练习 113 答案：避免重复

```
##
# Read a collection of words entered by the user. Display each word entered by the user only
# once, in the same order that the words were entered.
#

# Read words from the user and store them in a list
words = []
word = input("Enter a word (blank line to quit): ")
while word != "":
  # Only add the word to the list if
  # it is not already present in it
  if word not in words:
    words.append(word)

  # Read the next word from the user
  word = input("Enter a word (blank line to quit): ")

# Display the unique words
for word in words:
  print(word)
```

> 表达式 word not in words 和 not(word in words)是等价的。

练习 114 答案：负数、零和正数

```
##
# Read a collection of integers from the user. Display all of the negative numbers, followed
# by all of the zeros, followed by all of the positive numbers.
#

# Create three lists to store the negative, zero and positive values
negatives = []
zeros = []
positives = []
```

> 这个解使用一个列表来跟踪输入的零。但是，因为所有的零都是相同的，所以实际上没必要保存它们。相反，可使用一个整数变量来计算零的数量，然后在程序的后面显示许多零。

```
# Read all of the integers from the user, storing each integer in the correct list
line = input("Enter an integer (blank to quit): ")
while line != "":
  num = int(line)
```

```
    if num < 0:
      negatives.append(num)
    elif num > 0:
      positives.append(num)
    else:
      zeros.append(num)

    # Read the next line of input from the user
    line = input("Enter an integer (blank to quit): ")

# Display all of the negative values, then all of the zeros, then all of the positive values
print("The numbers were: ")

for n in negatives:
  print(n)

for n in zeros:
  print(n)

for n in positives:
  print(n)
```

练习 116 答案：完全数

```
##
# A number, n, is a perfect number if the sum of the proper divisors of n is equal to n. This
# program displays all of the perfect numbers between 1 and LIMIT.
#
from proper_divisors import properDivisors

LIMIT = 10000
## Determine whether or not a number is perfect. A number is perfect if the sum of its proper
# divisors is equal to the number itself.
# @param n the number to check for perfection
# @return True if the number is perfect, False otherwise
def isPerfect(n):
  # Get a list of the proper divisors of n
  divisors = properDivisors(n)

  # Compute the total of all of the divisors
  total = 0
  for d in divisors:
    total = total + d

  # Determine whether or not the number is perfect and return the appropriate result
  if total == n:
    return True
  return False
```

> total 也可使用 Python 的内置 sum() 函数来计算。这将把 total 的计算简化为一行。

```
# Display all of the perfect numbers between 1 and LIMIT
def main():
  print("The perfect numbers between 1 and", LIMIT, "are:")
  for i in range(1, LIMIT + 1):
    if isPerfect(i):
      print("  ", i)

# Call the main function
main()
```

练习 120 答案：格式化列表

```
##
# Display a list of items so that they are separated by commas and the word ''and'' appears
# between the final two items.
#

## Format a list of items so that they are separated by commas and ''and''
# @param items the list of items to format
# @return a string containing the items with the desired formatting
def formatList(items):
  # Handle lists of 0 and 1 items as special cases
  if len(items) == 0:
    return "<empty>"
  if len(items) == 1:
    return str(items[0])

  # Loop over all of the items in the list except the last two
  result = ""
  for i in range(0, len(items) - 2):
    result = result + str(items[i]) + ", "
```

将每个项连接到结果之前，通过调用 str 函数将其显式转换为字符串。这允许 formatList 函数格式化包括字符串和数字的列表。

```
  # Add the second last and last items to the result, separated by ''and''
  result = result + str(items[len(items) - 2]) + " and "
  result = result + str(items[len(items) - 1])
  # Return the result
  return result

# Read several items entered by the user and display them with nice formatting
def main():
  # Read items from the user until a blank line is entered
  items = []
  line = input("Enter an item (blank to quit): ")
  while line != "":
```

```
        items.append(line)
        line = input("Enter an item (blank to quit): ")

    # Format and display the items
    print("The items are %s." % formatList(items))

# Call the main function
main()
```

练习 121 答案：随机的彩票号码

```
##
# Compute random but distinct numbers for a lottery ticket.
#
from random import randrange

MIN_NUM = 1
MAX_NUM = 49
NUM_NUMS = 6
```

> 使用常量可很容易地为其他彩票重新配置程序。

```
# Use a list to store the numbers on the ticket
ticket_nums = []

# Generate NUM_NUMS random but distinct numbers
for i in range(NUM_NUMS):
    # Generate a number that isn't already on the ticket
    rand = randrange(MIN_NUM, MAX_NUM + 1)
    while rand in ticket_nums:
        rand = randrange(MIN_NUM, MAX_NUM + 1)

    # Add the number to the ticket
    ticket_nums.append(rand)

# Sort the numbers into ascending order and display them
ticket_nums.sort()
print("Your numbers are: ", end="")
for n in ticket_nums:
    print(n, end=" ")
print()
```

练习 125 答案：洗牌

```
##
# Create a deck of cards and shuffle it.
#
from random import randrange
```

```python
## Construct a standard deck of cards with 4 suits and 13 values per suit
# @return a list of cards, with each card represented by two characters
def createDeck():
    # Create a list to hold the cards
    cards = []
    # For each suit and each value
    for suit in ["s", "h", "d", "c"]:
        for value in ["2", "3", "4", "5", "6", "7", "8", "9", \
                      "T", "J", "Q", "K", "A"]:
            # Construct the card and add it to the list
            cards.append(value + suit)

    # Return the complete deck of cards
    return cards

## Shuffle a deck of cards by modifying the deck passed to the function
# @param cards the list of cards to shuffle
# @return (None)
def shuffle(cards):
    # For each card
    for i in range(0, len(cards)):
        # Pick a random index between the current index and the end of the list
        other_pos = randrange(i, len(cards))

        # Swap the current card with the one at the random position
        temp = cards[i]
        cards[i] = cards[other_pos]
        cards[other_pos] = temp

# Display a deck of cards before and after it has been shuffled
def main():
    cards = createDeck()
    print("The original deck of cards is: ")
    print(cards)
    print()

    shuffle(cards)
    print("The shuffled deck of cards is: ")
    print(cards)

    # Call the main function only if this file has not been imported into another program
if __name__ == "__main__":
    main(
```

练习 128 答案：确定元素个数

```python
##
# Count the number of elements in a list that are greater than or equal to some minimum
# value and less than some maximum value.
#

## Determine how many elements in data are greater than or equal to mn and less than mx
# @param data the list of values to examine
# @param mn the minimum acceptable value
# @param mx the exclusive upper bound on acceptability
# @return the number of elements, e, such that mn <= e < mx
def countRange(data, mn, mx):
  # Count the number of elements within the acceptable range
  count = 0
  for e in data:
    # Check each element
    if mn <= e and e < mx:
      count = count + 1

  # Return the result
  return count

# Demonstrate the countRange function
def main():
  data = [1, 2, 3, 4, 5, 6, 7, 8, 9, 10]

  # Test a case where some elements are within the range
  print("Counting the elements in [1..10] between 5 and 7...")
  print("Result: %d Expected: 2" % countRange(data, 5, 7))

  # Test a case where all elements are within the range
  print("Counting the elements in [1..10] between -5 and 77...")
  print("Result: %d Expected: 10" % countRange(data, -5, 77))

  # Test a case where no elements are within the range
  print("Counting the elements in [1..10] between 12 and 17...")
  print("Result: %d Expected: 0" % countRange(data, 12, 17))

  # Test a case where the list is empty
  print("Counting the elements in [] between 0 and 100...")
  print("Result: %d Expected: 0" % countRange([], 0, 100))

  # Test a case with duplicate values
  data = [1, 2, 3, 4, 1, 2, 3, 4]
  print("Counting the elements in", data, "between 2 and 4...")
  print("Result: %d Expected: 4" % countRange(data, 2, 4))
```

```
# Call the main program
main()
```

练习 129 答案：标记字符串

```
##
# Tokenize a string containing a mathematical expression.
#

## Convert a mathematical expression into a list of tokens
# @param s the string to tokenize
# @return a list of the tokens in s, or an empty list if an error occurs
def tokenList(s) :
    # Remove all of the spaces from s
    s = s.replace(" ", "")

    # Loop through all of the characters in the string, identifying the tokens and adding them to
    # the list
    tokens = []
    i = 0
    while i < len(s):
        # Handle the tokens that are always a single character: *, /, ^, ( and )
        if s[i] == "*" or s[i] == "/" or s[i] == "^" or \
           s[i] == "(" or s[i] == ")" or s[i] == "+" or s[i] == "-":
            tokens.append(s[i])
            i = i + 1
        # Handle a number without a leading + or -
        elif s[i] >= "0" and s[i] <= "9":
            num = ""
            # Keep on adding characters to the token as long as they are digits
            while i < len(s) and s[i] >= "0" and s[i] <= "9":
                num = num + s[i]
                i = i + 1
            tokens.append(num)

        # Any other character means the expression is not valid. Return an empty list to indicate
        # that an error occurred.
        else:
            return []

    return tokens

# Read an expression from the user, tokenize it, and display the result
def main():
    exp = input("Enter a mathematical expression: ")
    tokens = tokenList(exp)
    print("The tokens are:", tokens)
```

```python
# Call the main function only if this file has not been imported into another program
if __name__ == "__main__":
    main()
```

练习 130 答案：一元运算符和二元运算符

```python
##
# Differentiate between unary and binary + and - operators.
#
from token_list import tokenList

## Identify occurrences of unary + and - operators within a list of tokens and replace them
# with u+ and u- respectively
# @param tokens a list of tokens that may include unary + and - operators
# @return a list of tokens where unary + and - operators have been replaced with u+ and u-
Def identifyUnary(tokens):
    retval = []

    # Process each token in the list
    for i in range(len(tokens)):
        # If the first token in the list is + or - then it is a unary operator
        if i == 0 and (tokens[i] == "+" or tokens[i] == "-"):
            retval.append("u" + tokens[i])
        # If the token is a + or - and the previous token is an operator or an open parenthesis
        # then it is a unary operator
        elif i > 0 and (tokens[i] == "+" or tokens[i] == "-") and \
            (tokens[i-1] == "+" or tokens[i-1] == "-" or
             tokens[i-1] == "*" or tokens[i-1] == "/" or
             tokens[i-1] == "("):
            retval.append("u" + tokens[i])
        # Any other token is not a unary operator so it is appended to the result without modification
        else:
            retval.append(tokens[i])

    # Return the new list of tokens where the unary operators have been marked
    return retval

# Demonstrate that unary operators are marked correctly
def main():
    # Read an expression from the user, tokenize it, and display the result
    exp = input("Enter a mathematical expression: ")
    tokens = tokenList(exp)
    print("The tokens are:", tokens)

    # Identify the unary operators in the list of tokens
    marked = identifyUnary(tokens)
    print("With unary operators marked: ", marked)
```

```
# Call the main function only if this file has not been imported into another program
if __name__ == "__main__":
    main()
```

练习 134 答案：生成列表的所有子列表

```
##
# Compute all of the sublists of a list.
#
## Generate a list of of all of the sublists of a list
# @param data the list for which the sublists are generated
# @return a list containing all of the sublists of data
def allSublists(data):
    # Start out with the empty list as the only sublist of data
    sublists = [[]]

    # Generate all of the sublists of data from length 1 to len(data)
    for length in range(1, len(data) + 1):
        # Generate the sublists starting at each index
        for i in range(0, len(data) - length + 1):
            # Add the current sublist to the list of sublists
            sublists.append(data[i : i + length])

    # Return the result
    return sublists

# Demonstrate the allSublists function
def main():
    print("The sublists of [] are: ")
    print(allSublists([]))

    print("The sublists of [1] are: ")
    print(allSublists([1]))

    print("The sublists of [1, 2] are: ")
    print(allSublists([1, 2]))

    print("The sublists of [1, 2, 3] are: ")
    print(allSublists([1, 2, 3]))

    print("The sublists of [1, 2, 3, 4] are: ")
    print(allSublists([1, 2, 3, 4]))

# Call the main function
main()
```

> 包含空列表的列表用[[]]表示。

练习 135 答案：埃拉托色尼筛法

```
##
# Identify all of the prime numbers from 2 to some limit entered by the user using the
# Sieve of Eratosthenes.
#
# Read the limit from the user
limit = int(input("Identify all primes up to what limit? "))

# Create a list that contains all of the integers from 0 to limit
nums = []
for i in range(0, limit + 1):
  nums.append(i)

# "Cross out" 1 by replacing it with a 0
nums[1] = 0
# "Cross out" all of the multiples of each prime number that we discover
p = 2
while p < limit:
  # "Cross out" all multiples of p (but not p itself)
  for i in range(p*2, limit + 1, p):
    nums[i] = 0

  # Find the next number that is not "crossed out"
  p = p + 1
  while p < limit and nums[p] == 0:
    p = p + 1

# Display the result
print("The primes up to", limit, "are:")
for i in nums:
  if nums[i] != 0:
    print(i)
```

第14章 "字典"练习答案

练习 136 答案：反向查找

```
##
# Conduct a reverse lookup on a dictionary, finding all of the keys that map to the provided
# value.
#

## Conduct a reverse lookup on a dictionary
# @param data the dictionary on which the reverse lookup is performed
# @param value the value to search for in the dictionary
# @return a list (possibly empty) of keys from data that map to value
def reverseLookup(data, value):
  # Construct a list of the keys that map to value
  keys = []

  # Check each key and add it to keys if the values match
  for key in data:
    if data[key] == value:
      keys.append(key)

  # Return the list of keys
  return keys
```

> 字典中的每个键必须是唯一的。但是，值可能会重复。因此，执行反向查找可能会使零个、一个或多个键与提供的值匹配。

```
# Demonstrate the reverseLookup function
def main():
    # A dictionary mapping 4 French words to their English equivalents
    frEn = {"le" : "the", "la" : "the", "livre" : "book", \
            "pomme" : "apple"}

    # Demonstrate the reverseLookup function with 3 cases: One that returns multiple keys,
    # one that returns one key, and one that returns no keys
    print("The french words for 'the' are: ", \
          reverseLookup(frEn, "the"))
    print("Expected: ['le', 'la']")
    print()
    print("The french word for 'apple' is: ", \
          reverseLookup(frEn, "apple"))
    print("Expected: ['pomme']")
    print()
    print("The french word for 'asdf' is: ", \
          reverseLookup(frEn, "asdf"))
    print("Expected: []")

# Call the main function only if this file has not been imported into another program
if __name__ == "__main__":
    main()
```

练习 137 答案：两个骰子的模拟

```
##
# Simulate rolling two dice many times and compare the simulated results to the results
# expected by probability theory.
#
from random import randrange

NUM_RUNS = 1000
D_MAX = 6

## Simulate rolling two six-sided dice
# @return the total from rolling two simulated dice
def twoDice():
    # Simulate two dice
    d1 = randrange(1, D_MAX + 1)
    d2 = randrange(1, D_MAX + 1)

    # Return the total
    return d1 + d2

# Simulate many rolls and display the result
def main():
    # Create a dictionary of expected proportions
```

第 14 章 "字典" 练习答案

```
expected = {2: 1/36, 3: 2/36, 4: 3/36, 5: 4/36, 6: 5/36, \
            7: 6/36, 8: 5/36, 9: 4/36, 10: 3/36, \
            11: 2/36, 12: 1/36}

# Create a dictionary that maps from the total of two dice to the number of occurrences
counts = {2: 0, 3: 0, 4: 0, 5: 0, 6: 0, 7: 0, \
          8: 0, 9: 0, 10: 0, 11: 0, 12: 0}
```

初始化每个字典，因此它的键是 2、3、4、5、6、7、8、9、10、11 和 12。在预期的字典中，该值初始化为当掷两个 6 面骰子时每个键发生的概率。在 counts 字典中，每个值都初始化为 0。counts 的值随着模拟的运行而增加。

```
# Simulate NUM RUNS rolls, and count each roll
for i in range(NUM_RUNS):
    t = twoDice()
    counts[t] = counts[t] + 1

# Display the simulated proportions and the expected proportions
print("Total Simulated Expected")
print("      Percent Percent")
for i in sorted(counts.keys()):
    print("%5d %11.2f %8.2f" % \
          (i, counts[i] / NUM_RUNS * 100, expected[i] * 100))

# Call the main function
main()
```

练习 142 答案：独特的字符

```
##
# Compute the number of unique characters in a string using a dictionary.
#

# Read the string from the user
s = input("Enter a string: ")

# Add each character to a dictionary with a value of True. Once we are done the number
# of keys in the dictionary will be the number of unique characters in the string.
characters = {}
for ch in s:
    characters[ch] = True
```

字典中的每个键都必须有一个与之关联的值。但在这个解中，这个值从未用过。因此，这里选择将 True 与每个键关联，但可以使用任何其他值而不是 True。

```
# Display the result
print("That string contained", len(characters), \
      "unique character(s).")
```

len 函数返回字典中的键数。

练习 143 答案：字谜

```
##
# Determine whether or not two strings are anagrams and report the result.
#

## Compute the frequency distribution for the characters in a string
# @param s the string to process
# @return a dictionary mapping each character to its count
def characterCounts(s):
    # Create a new, empty dictionary
    counts = {}

    # Update the count for each character in the string
    for ch in s:
        if ch in counts:
            counts[ch] = counts[ch] + 1
        else:
            counts[ch] = 1

    # Return the result
    return counts

# Determine if two strings entered by the user are anagrams
def main():
    # Read the strings from the user
    s1 = input("Enter the first string: ")
    s2 = input("Enter the second string: ")

    # Get the character counts for each string
    counts1 = characterCounts(s1)
    counts2 = characterCounts(s2)

    # Display the result
    if counts1 == counts2:
        print("Those strings are anagrams.")
    else:
        print("Those strings are not anagrams.")
```

第 14 章 "字典"练习答案

当且仅当两个字典具有相同的键,且对于每个键 k,两个字典中与 k 关联的值相同时,两个字典才是相等的。

```
# Call the main function
main()
```

练习 145 答案:Scrabble 分

```
##
# Use a dictionary to compute the Scrabble™ score for a word.
#

# Initialize the dictionary so that it maps from letters to points
points = {"A": 1, "B": 3, "C": 3, "D": 2, "E": 1, "F": 4, \
          "G": 2, "H": 4, "I": 1, "J": 2, "K": 5, "L": 1, \
          "M": 3, "N": 1, "O": 1, "P": 3, "Q": 10, "R": 1, \
          "S": 1, "T": 1, "U": 1, "V": 4, "W": 4, "X": 8, \
          "Y": 4, "Z": 10}

# Read a word from the user
word = input("Enter a word: ")

# Compute the score for the word
uppercase = word.upper()
score = 0
for ch in uppercase:
    score = score + points[ch]

# Display the result
print(word, "is worth", score, "points.")
```

当用户以大写、混合或小写形式输入单词时,将该单词转换为大写,从而计算出正确结果。也可通过将所有小写字母添加到字典中来实现。

练习 146 答案:制作一张宾果卡

```
##
# Create and display a random Bingo card.
#
from random import randrange

NUMS_PER_LETTER = 15

## Create a Bingo card with randomly generated numbers
# @return a dictionary representing the card where the keys are the strings ''B'',
  ''I'', ''N'',
#         ''G'', and ''O'', and the values are lists of the numbers that appear under each letter
#         from top to bottom
```

```python
def createCard():
    card = {}

    # The range of integers that can be generated for the current letter
    lower = 1
    upper = 1 + NUMS_PER_LETTER

    # For each of the five letters
    for letter in ["B", "I", "N", "G", "O"]:
        # Start with an empty list for the letter
        card[letter] = []

        # Keep generating random numbers until we have 5 unique ones
        while len(card[letter]) != 5:
            next_num = randrange(lower, upper)
            # Ensure that we do not include any duplicate numbers
            if next_num not in card[letter]:
                card[letter].append(next_num)

        # Update the range of values that will be generated for the next letter
        lower = lower + NUMS_PER_LETTER
        upper = upper + NUMS_PER_LETTER

    # Return the card
    return card

## Display a Bingo card with nice formatting
# @param card the Bingo card to display
# @return (None)
def displayCard(card):
    # Display the headings
    print("B  I  N  G  O")

    # Display the numbers on the card
    for i in range(5):
        for letter in ["B", "I", "N", "G", "O"]:
            print("%2d " % card[letter][i], end="")
        print()

# Create a random Bingo card and display it
def main():
    card = createCard()
    displayCard(card)

# Call the main function only if this file has not been imported into another program
if __name__ == "__main__":
    main()
```

第15章 "文件和异常"练习答案

练习 149 答案：显示文件的头部

```
##
# Display the head (first 10 lines) of a file whose name is provided as a command
line argument.
#
import sys

NUM_LINES = 10

# Verify that exactly one command line argument (in addition to the .py file) was supplied
if len(sys.argv) != 2:
    print("Provide the file name as a command line argument.")
    quit()

try:
    # Open the file for reading
    inf = open(sys.argv[1], "r")

    # Read the first line from the file
    line = inf.readline()

    # Keep looping until we have either read and displayed 10 lines or we have reached the end
    # of the file
```

> 调用 quit 函数时，程序立即结束。

```
    count = 0
    while count < NUM_LINES and line != "":
        # Remove the trailing newline character and count the line
        line = line.rstrip()
        count = count + 1

        # Display the line
        print(line)

        # Read the next line from the file
        line = inf.readline()

    # Close the file
    inf.close()

except IOError:
    # Display a message if something goes wrong while accessing the file
    print("An error occurred while accessing the file.")
```

练习 150 答案：显示文件的尾部

```
##
# Display the tail (last lines) of a file whose name is provided as a command line argument.
#
import sys

NUM_LINES = 10

# Verify that exactly one command line argument (in addition to the .py file) was provided
if len(sys.argv) != 2:
    print("Provide the file name as a command line argument.")
    quit()

try:
    # Open the file for reading
    inf = open(sys.argv[1], "r")

    # Read through the file, always saving the NUM_LINES most recent lines
    lines = []
    for line in inf:
        # Add the most recent line to the end of the list
        lines.append(line)
        # If we now have more than NUM_LINES lines then remove the oldest one
        if len(lines) > NUM_LINES:
            lines.pop(0)

    # Close the file
```

第 15 章 "文件和异常"练习答案

```
  inf.close()

except:
  print("An error occurred while processing the file.")
  quit()

# Display the last lines of the file
for line in lines:
  print(line, end="")
```

练习 151 答案：连接多个文件

```
##
# Concatenate one or more files and display the result.
#
import sys

# Ensure that at least one command line argument (in addition to the .py file) has been provided
if len(sys.argv) == 1:
  print("You must provide at least one file name.")
  quit()

# Process all of the files provided on the command line
for i in range(1, len(sys.argv)):
  fname = sys.argv[i]
  try:
    # Open the current file for reading
    inf = open(fname, "r")

    # Display the file
    for line in inf:
      print(line, end="")

    # Close the file
    inf.close()

  except:
    # Display a message, but do not quit, so that the program will go on and process any
    # subsequent files
    print("Couldn't open/display", fname)
```

> sys.argv 中位置 0 处的元素是正在执行的 Python 文件。因此，for 循环从列表中的第 1 个位置开始处理文件名。

练习 156 答案：对一组数字求和

```
##
# Compute the sum of numbers entered by the user, ignoring non-numeric input.
#
```

```
# Read the first line of input from the user
line = input("Enter a number: ")
total = 0

# Keep reading until the user enters a blank line
while line != "":
  try:
    # Try and convert the line to a number
    num = float(line)
    # If the conversion succeeds then add it to the total and display it
    total = total + num
    print("The total is now", total)

  except ValueError:
    # Display an error message before going on to read the next value
    print("That wasn't a number.")

  # Read the next number
  line = input("Enter a number: ")

# Display the total
print("The grand total is", total)
```

练习 158 答案：删除注释

```
##
# Remove all of the comments from a Python file (ignoring the case where a comment
# character occurs within a string)
#
# Read the file name and open the input file
try:
  in_name = input("Enter the name of a Python file: ")
  inf = open(in_name, "r")

except:
  # Display an error message and quit if the file was not opened successfully
  print("A problem was encountered with the input file.")
  print("Quitting...")
  quit()

# Read the file name and open the output file
try:
  out_name = input("Enter the output file name: ")
  outf = open(out_name, "w")

except:
  # Close the input file, display an error message and quit if the file was not opened
  # successfully
```

```
        inf.close()
        print("A problem was encountered with the output file.")
        print("Quitting...")
        quit()
try:
    # Read all of the lines from the input file, remove the comments from them, and save the
    # modified lines to a new file
    for line in inf:
        # Find the position of the comment character (-1 if there isn't one)
        pos = line.find("#")

        # If there is a comment then create a slice of the string that excludes it and store it back
        # into line
        if pos > -1:
            line = line[0 : pos]
            line = line + "\n"

        # Write the (potentially modified) line to the file
        outf.write(line)

    # Close the files
    inf.close()
    outf.close()

except:
    # Display an error message if something went wrong while processing the file
    print("A problem was encountered while processing the file.")
    print("Quitting...")
```

> 注释字符的位置存储在 pos 中。因此，line[0:pos]是从头到注释字符(不含)的所有字符。

练习 159 答案：两个单词的随机密码

```
##
# Generate a password by concatenating two random words. The password will be between
# 8 and 10 letters, and each word will be at least three letters long.
#
from random import randrange

WORD_FILE = "../Data/words.txt"
```

> 我们创建的密码是 8、9 或 10 个字母。因为可接受的最短单词是 3 个字母，密码必须包含 2 个单词，所以密码不能包含超过 7 个字母的单词。

```
# Read all of the words from the file, only keeping those between 3 and 7 letters in length,
# and store them in a list
words = []
inf = open(WORD_FILE, "r")
```

```
for line in inf:
  # Remove the newline character
  line = line.rstrip()

  # Keep words that are between 3 and 7 letters long
  if len(line) >= 3 and len(line) <= 7:
    words.append(line)

# Close the file
inf.close()

# Randomly select the first word for the password. It can be any word.
first = words[randrange(0, len(words))]
first = first.capitalize()

# Keep selecting a second word until we find one that doesn't make the password too short
# or too long
password = first
while len(password) < 8 or len(password) > 10:
  second = words[randrange(0, len(words))]
  second = second.capitalize()
  password = first + second

# Display the random password
print("The random password is:", password)
```

练习 162 答案：一本没有 E 的书...

```
# Determine and display the proportion of words that include each letter of the alphabet. The
# letter that is used in the smallest proportion of words is highlighted at the end of the
# program's output.
#
WORD_FILE = "../Data/words.txt"

# Create a dictionary that counts the number of words containing each letter. Initialize the
# count for each letter to 0.
counts = {}
for ch in "ABCDEFGHIJKLMNOPQRSTUVWXYZ":
  counts[ch] = 0

# Open the file, process each word, and update the counts dictionary
num_words = 0
inf = open(WORD_FILE, "r")
for word in inf:
  # Convert the word to uppercase and remove the newline character
  word = word.upper().rstrip()
  # Before we can update the dictionary we need to generate a list of the unique letters in the
  # word. Otherwise we will increase the count multiple times for words that contain repeated
```

```
  # letters. We also need to ignore any non-letter characters that might be present.
  unique = []
  for ch in word:
    if ch not in unique and ch >= "A" and ch <= "Z":
      unique.append(ch)

  # Now increment the counts for all of the letters that are in the list of unique characters
  for ch in unique:
    counts[ch] = counts[ch] + 1

  # Keep track of the number of words that we have processed
  num_words = num_words + 1

# Close the file
inf.close()

# Display the result for each letter. While displaying the results we will also determine which
# character had the smallest count so that we can display it again at the end of the program.
smallest_count = min(counts.values())
for ch in sorted(counts):
  if counts[ch] == smallest_count:
    smallest_letter = ch
  percentage = counts[ch] / num_words * 100
  print(ch, "occurs in %.2f percent of words" % percentage)

# Display the letter that is easiest to avoid based on the number of words in which it appears
print()
print("The letter that is easiest to avoid is", smallest_letter)
```

练习 163 答案：排名第一的名字

```
##
# Display all of the girls' and boys' names that were the most popular in at least one year
# between 1900 and 2012.
#
FIRST_YEAR = 1900
LAST_YEAR = 2012

## Load the first line from the file, extract the name, and add it to the names list if it is not
# already present.
# @param fname the name of the file from which the data will be read
# @param names the list to add the name to (if it isn't already present)
# @return (None)
def LoadAndAdd(fname, names):
  # Open the file, read the first line, and extract the name
  inf = open(fname, "r")
  line = inf.readline()
```

```
    inf.close()
    parts = line.split()
    name = parts[0]

    # Add the name to the list if it is not already present
    if name not in names:
        names.append(name)

# Display the girls' and boys' names that reached #1 in at least one year between 1900 and 2012
def main():
    # Create two lists to store the most popular names
    girls = []
    boys = []

    # Process each year in the range, reading the first line out of the girl file and the boy file
    for year in range(FIRST_YEAR, LAST_YEAR + 1):
        girl_fname = "../Data/BabyNames/" + str(year) + \
                     "_GirlsNames.txt"
        boy_fname = "../Data/BabyNames/" + str(year) + \
                    "_BoysNames.txt"
```

> 本题的解假设婴儿名数据文件存储在与 Python 程序不同的文件夹中。如果数据文件和程序在同一文件夹中，就应该省略 ../Data/BabyNames/。

```
        LoadAndAdd(girl_fname, girls)
        LoadAndAdd(boy_fname, boys)

    # Display the lists
    print("Girls' names that reached #1:")
    for name in girls:
        print("  ", name)
    print()

    print("Boys' names that reached #1: ")
    for name in boys:
        print("  ", name)

# Call the main function
main()
```

练习 167 答案：拼写检查

```
##
# Find and list all of the words in a file that are misspelled.
#
from only_words import onlyTheWords
import sys
```

```
WORDS_FILE = "../Data/words.txt"

# Ensure that the program has the correct number of command line arguments
if len(sys.argv) != 2:
    print("One command line argument must be provided.")
    print("Quitting...")
    quit()

# Open the file. Quit if the file is not opened successfully.
try:
    inf = open(sys.argv[1], "r")
except:
    print("Failed to open '%s' for reading. Quitting..." % \
          sys.argv[1])
    quit()

# Load all of the words into a dictionary of valid words. The  value 0 is associated with each
# word, but it is never used.
valid = {}
words_file = open(WORDS_FILE, "r")

for word in words_file:
    # Convert the word to lowercase and remove the trailing newline character
    word = word.lower().rstrip()

    # Add the word to the dictionary
    valid[word] = 0
words_file.close()
```

> 此解使用字典，其中键是有效单词，但字典中的值从未用过。因此，如果熟悉集合，那么该数据结构是更好的选择。不使用列表，因为检查键是否在字典中比检查元素是否在列表中要快。

```
# Read each line from the file, adding any misspelled words to the list of misspellings
misspelled = []
for line in inf:
    # Discard the punctuation marks by calling the function developed in Exercise 117
    words = onlyTheWords(line)
    for word in words:
        # Only add to the misspelled list if the word is misspelled and not already in the list
        if word.lower() not in valid and word not in misspelled:
            misspelled.append(word)

# Close the file being checked
inf.close()

# Display the misspelled words, or a message indicating that no words are misspelled
```

```
if len(misspelled) == 0:
  print("No words were misspelled.")
else:
  print("The following words are misspelled:")
  for word in misspelled:
    print(" ", word)
```

练习 169 答案：编辑文件中的文本

```
##
# Redact a text file by removing all occurrences of sensitive words. The redacted version
# of the text is written to a new file.
#
# Note that this program does not perform any error checking, and it does not implement
# case insensitive redaction.
#

# Get the name of the input file and open it
inf_name = input("Enter the name of the text file to redact: ")
inf = open(inf_name, "r")

# Get the name of the sensitive words file and open it
sen_name = input("Enter the name of the sensitive words file: ")
sen = open(sen_name, "r")

# Load all of the sensitive words into a list
words = []
line = sen.readline()
while line != "":
  line = line.rstrip()
  words.append(line)

  line = sen.readline()

# Close the sensitive words file
sen.close()
```

> 此时可关闭敏感词文件，因为所有单词都已读入一个列表。不需要从该文件中读取其他数据。

```
# Get the name of the output file and open it
outf_name = input("Enter the name for the new redacted file: ")
outf = open(outf_name, "w")

# Read each line from the input file. Replace all of the sensitive words with asterisks. Then
# write the line to the output file.
line = inf.readline()
```

```
while line != "":
    # Check for and replace each sensitive word. The number of asterisks matches the number
    # of letters in the word being redacted.
    for word in words:
        line = line.replace(word, "*" * len(word))

    # Write the modified line to the output file
    outf.write(line)

    # Read the next line from the input file
    line = inf.readline()

# Close the input and output files
inf.close()
outf.close()
```

练习 170 答案：缺少注释

```
##
# Find and display the names of Python functions that are not immediately preceded by a
# comment.
#
from sys import argv
# Verify that at least one file name has been provided as a command line argument
if len(argv) == 1:
    print("At least one filename must be provided as a", \
          "command line argument.")
    print("Quitting...")
    quit()

# Process each file provided as a command line argument
for fname in argv[1 : len(argv)]:
    # Attempt to process the file
    try:
        inf = open(fname, "r")

        # As we move through the file we need to keep a copy of the previous line so that we
        # can check to see if it starts with a comment character. We also need to keep track
        # of the line number within the file.
        prev = " "
        lnum = 1
```

必须将 prev 变量初始化为长度至少为一个字符的字符串。否则，当文件中被检查的第一行是函数定义时，程序将崩溃。

```
        # Check each line in the current file
        for line in inf:
```

```
        # If the line is a function definition and the previous line is not a comment
        if line.startswith("def ") and prev[0] != "#":
            # Find the first ( on the line so that we know where the function name ends
            bracket_pos = line.index("(")
            name = line[4 : bracket_pos]

            # Display information about the function that is missing its comment
            print("%s line %d: %s" % (fname, lnum, name))

        # Save the current line and update the line counter
        prev = line
        lnum = lnum + 1

    # Close the current file
    inf.close()

except:
    print("A problem was encountered with file '%s'." % fname)
    print("Moving on to the next file...")
```

第16章 "递归"练习答案

练习 173 答案：把这些值加起来

```python
##
# Total a collection of numbers entered by the user. The user will enter a blank line to
# indicate that no further numbers will be entered and the total should be displayed.
#

## Total all of the numbers entered by the user until the user enters a blank line
# @return the total of the entered values
def readAndTotal():
    # Read a value from the user
    line = input("Enter a number (blank to quit): ")

    # Base case: The user entered a blank line so the total is 0
    if line == "":
        return 0
    else:
        # Recursive case: Convert the current line to a number and use recursion to read the
        # subsequent lines
        return float(line) + readAndTotal()

# Read a collection of numbers from the user and display the total
def main():
    # Read the values from the user and compute the total
```

```
    total = readAndTotal()

    # Display the total
    print("The total of all those values is", total)

# Call the main function
main()
```

练习 178 答案：递归回文

```
##
# Determine whether or not a string entered by the user is a palindrome using recursion.
#

## Determine whether or not a string is a palindrome
# @param s the string to check
# @return True if the string is a palindrome, False otherwise
def isPalindrome(s):
    # Base case: The empty string is a palindrome. So is a string containing only 1 character.
    if len(s) <= 1:
        return True

    # Recursive case: The string is a palindrome only if the first and last characters match, and
    # the rest of the string is a palindrome
    return s[0] == s[len(s) - 1] and \
           isPalindrome(s[1 : len(s) - 1])

# Check whether or not a string entered by the user is a palindrome
def main():
    # Read the string from the user
    line = input("Enter a string: ")

    # Check its status and display the result
    if isPalindrome(line):
        print("That was a palindrome!")
    else:
        print("That is not a palindrome.")

# Call the main function
main()
```

练习 180 答案：字符串编辑距离

```
##
# Compute and display the edit distance between two strings.
#

## Compute the edit distance between two strings. The edit distance is the minimum number of
```

```
# insert, delete and substitute operations needed to transform one string into the other.
# @param s the first string
# @param t the second string
# @return the edit distance between the strings
def editDistance(s, t):
    # If one string is empty, then the edit distance is one insert operation for each letter in the
    # other string
    if len(s) == 0:
        return len(t)
    elif len(t) == 0:
        return len(s)
    else:
        cost = 0
        # If the last characters are not equal then set cost to 1
        if s[len(s) - 1] != t[len(t) - 1]:
            cost = 1

        # Compute three distances
        d1 = editDistance(s[0 : len(s) - 1], t) + 1
        d2 = editDistance(s, t[0 : len(t) - 1]) + 1
        d3 = editDistance(s[0 : len(s) - 1] , t[0 : len(t) - 1]) + \
            cost

        # Return the minimum distance
        return min(d1, d2, d3)

# Compute the edit distance between two strings entered by the user
def main():
    # Read two strings from the user
    s1 = input("Enter a string: ")
    s2 = input("Enter another string: ")

    # Compute and display the edit distance
    print("The edit distance between %s and %s is %d." % \
        (s1, s2, editDistance(s1, s2)))

# Call the main function
main()
```

练习 183 答案：元素序列

```
##
# Identify the longest sequence of elements that can follow an element entered by the
# user where each element in the sequence begins with the same letter as the last letter
# of its predecessor.
#
ELEMENTS_FNAME = "../Data/Elements.csv"
```

```python
## Find the longest sequence of words, beginning with start, where each word begins with
#  the last letter of its predecessor
#  @param start the first word in the sequence
#  @param words the list of words that can occur in the sequence
#  @return the longest list of words beginning with start that meets the constraints
#          outlined previously
def longestSequence(start, words):
  # Base case: If start is empty then the list of words is empty
  if start == "":
    return []

  # Find the best (longest) list of words by checking each possible word that could appear
  # next in the sequence
  best = []
  last_letter = start[len(start) - 1].lower()
  for i in range(0, len(words)):
    first_letter = words[i][0].lower()

    # If the first letter of the next word matches the last letter of the previous word
    if first_letter == last_letter:
      # Use recursion to find a candidate sequence of words
      candidate = longestSequence(words[i], \
                  words[0 : i] + words[i + 1 : len(words)])
      # Save the candidate if it is better than the best sequence that we have seen previously
      if len(candidate) > len(best):
        best = candidate

  # Return the best candidate sequence, preceded by the starting word
  return [start] + best

## Load the names of all of the elements from the elements file
#  @return a list of all the element names
def loadNames():
  names = []

  # Open the element data set
  inf = open(ELEMENTS_FNAME, "r")

  # Load each line, storing the element name in a list
  for line in inf:
    line = line.rstrip()
    parts = line.split(",")
    names.append(parts[2])

  # Close the file and return the list
  inf.close()
  return names
```

```python
# Display a longest sequence of elements starting with an element entered by the user
def main():
    # Load all of the element names
    names = loadNames()

    # Read the starting element and capitalize it
    start = input("Enter the name of an element: ")
    start = start.capitalize()

    # Verify that the value entered by the user is an element
    if start in names:
        # Remove the starting element from the list of elements
        names.remove(start)

        # Find a longest sequence that begins with the starting element
        sequence = longestSequence(start, names)

        # Display the sequence of elements
        print("A longest sequence that starts with", start, "is:")
        for element in sequence:
            print(" ", element)
    else:
        print("Sorry, that wasn't a valid element name.")

# Call the main function
main()
```

练习 184 答案：把一张清单压平

```python
##
# Use recursion to flatten a list that may contain nested lists.
#

## Flatten a list so that all nested lists are removed
# @param data the list to flatten
# @return a flattened version of data
def flatten(data):
    # If data is empty then there is no work to do
    if data == []:
        return []

    # If the first element in data is a list
    if type(data[0]) == list:
        # Flatten the first element and flatten the remaining elements in the list
        return flatten(data[0]) + flatten(data[1:])
    else:
        # The first element in data is not a list so only the remaining elements in the list need
        # to be flattened
```

```
    return [data[0]] + flatten(data[1:])

# Demonstrate the flatten function
def main():
  print(flatten([[1, 2, 3], [4, [5, [6, 7]]], [[[8], 9], [10]]]))
  print(flatten([1, [2, [3, [4, [5, [6, [7, [8, [9, \
                [10]]]]]]]]]]))
  print(flatten([[[[[[[[[[1], 2], 3], 4], 5], 6], 7], 8], 9], \
                10]))
  print(flatten([1, 2, 3, 4, 5, 6, 7, 8, 9, 10]))
  print(flatten([]))

# Call the main function
main()
```

练习 186 答案：运行长度编码

```
# Perform run-length encoding on a string or a list using recursion.
#
## Perform run-length encoding on a string or a list
# @param data the string or list to encode
# @return a list where the elements at even positions are data values and the elements at odd
#         positions are counts of the number of times that the data value ahead of it should be
#         replicated
def encode(data):
  # If data is empty then no encoding work is necessary
  if len(data) == 0:
    return []
```

如果比较 data== ""，那么这个函数只对字符串有效。如果比较 data==[]，那么它只对列表有效。检查参数的长度可让函数同时处理字符串和列表。

```
  # Find the index of the first item that is not the same as the item at position 0 in data
  index = 1
  while index < len(data) and data[index] == data[index - 1]:
    index = index + 1

  # Encode the current character group
  current = [data[0], index]

  # Use recursion to process the characters from index to the end of the string
  return current + encode(data[index : len(data)])

# Demonstrate the encode function
def main():
  # Read a string from the user
  s = input("Enter some characters: ")
```

```
    # Display the result
    print("When those characters are run-length encoded," \
          "the result is:", encode(s))

# Call the main function
main()
```